NATURAL DISASTER RESEARCH, PREDICTION AND MITIGATION

FEDERAL FLOOD POLICY: BALANCING CHALLENGES AND LESSONS

NATURAL DISASTER RESEARCH, PREDICTION AND MITIGATION

Additional books in this series can be found on Nova's website
under the Series tab.

Additional E-books in this series can be found on Nova's website
under the E-books tab.

NATURAL DISASTER RESEARCH, PREDICTION AND MITIGATION

FEDERAL FLOOD POLICY: BALANCING CHALLENGES AND LESSONS

JAMES E. RYSANEK
EDITOR

Nova Science Publishers, Inc.
New York

Copyright © 2011 by Nova Science Publishers, Inc.

All rights reserved. No part of this book may be reproduced, stored in a retrieval system or transmitted in any form or by any means: electronic, electrostatic, magnetic, tape, mechanical photocopying, recording or otherwise without the written permission of the Publisher.

For permission to use material from this book please contact us:
Telephone 631-231-7269; Fax 631-231-8175
Web Site: http://www.novapublishers.com

NOTICE TO THE READER

The Publisher has taken reasonable care in the preparation of this book, but makes no expressed or implied warranty of any kind and assumes no responsibility for any errors or omissions. No liability is assumed for incidental or consequential damages in connection with or arising out of information contained in this book. The Publisher shall not be liable for any special, consequential, or exemplary damages resulting, in whole or in part, from the readers' use of, or reliance upon, this material. Any parts of this book based on government reports are so indicated and copyright is claimed for those parts to the extent applicable to compilations of such works.

Independent verification should be sought for any data, advice or recommendations contained in this book. In addition, no responsibility is assumed by the publisher for any injury and/or damage to persons or property arising from any methods, products, instructions, ideas or otherwise contained in this publication.

This publication is designed to provide accurate and authoritative information with regard to the subject matter covered herein. It is sold with the clear understanding that the Publisher is not engaged in rendering legal or any other professional services. If legal or any other expert assistance is required, the services of a competent person should be sought. FROM A DECLARATION OF PARTICIPANTS JOINTLY ADOPTED BY A COMMITTEE OF THE AMERICAN BAR ASSOCIATION AND A COMMITTEE OF PUBLISHERS.

Additional color graphics may be available in the e-book version of this book.

LIBRARY OF CONGRESS CATALOGING-IN-PUBLICATION DATA

Federal flood policy : balancing challenges and lessons / editor, James E. Rysanek.
    p. cm.
Includes index.
ISBN 978-1-61324-017-5 (hardcover)
 1. Flood control--Government policy--United States. I. Rysanek, James E.
TC423.F43 2011
 363.34'935610973--dc22
                           2011007122

Published by Nova Science Publishers, Inc. † New York

CONTENTS

Preface vii

Chapter 1 Federal Flood Policy Challenges:
Lessons from the 2008 Midwest Flood 1
Nicole T. Carter

Chapter 2 FEMA's Pre-Disaster Mitigation Program:
Overview and Issues 51
Francis X. McCarthy and Natalie Keegan

Chapter 3 Flood Risk Management and Levees:
A Federal Primer 83
Betsy A. Cody and Nicole T. Carter

Chapter 4 Midwest Flooding Disaster:
Rethinking Federal Flood Insurance? 97
Rawle O. King

Chapter 5 Midwest Floods of 2008:
Potential Impact on Agriculture 123
Randy Schnepf

Index 143

PREFACE

Floods remain a significant hazard in the United States. Developing and investing in flood-prone areas represents a tradeoff between the location's economic benefits and the exposure to a flood hazard. In the United States, flood mitigation, protection, emergency response and recovery roles and responsibilities are shared. Local governments are responsible for land use and zoning decisions that shape floodplain and coastal development. State and federal programs, policies, and investments influence community and individual decisions on managing flood risk. This new book examines federal flood policy challenges; FEMA's pre-disaster mitigation program; flood risk management and levees; federal flood insurance and the potential impact on agriculture.

Chapter 1- Floods remain a significant hazard in the United States. Developing and investing in flood-prone areas represents a tradeoff between the location's economic and other benefits and the exposure to a flood hazard. In the United States, flood mitigation, protection, emergency response, and recovery roles and responsibilities are shared. Local governments are responsible for land use and zoning decisions that shape floodplain and coastal development. State and federal programs, policies, and investments influence community and individual decisions on managing flood risk. The federal government constructs some of the nation's dams and levees, offers flood insurance, supports nonstructural risk reduction actions (known as hazard mitigation), and provides emergency response and disaster aid.

Chapter 2- Pre-Disaster Mitigation (PDM), as federal law and a program activity, began in 1997. Congress established a pilot program, which FEMA named "Project Impact," to test the concept of investing prior to disasters to reduce the vulnerability of communities to future disasters. P.L. 106-390, the

Disaster Mitigation Act of 2000, authorized the PDM program in law as Section 203 of the Robert T. Stafford Disaster Relief and Emergency Assistance Act.

From its beginnings as "Project Impact" to its current state, the PDM program has grown in its level of appropriated resources and the scope of participation nationwide. Along with that growth have come issues for Congressional consideration, including the approach for awarding grant funds, the eligibility of certain applicants, the eligibility of certain projects, the degree of commitment by state and local governments, and related questions.

Chapter 3- Midwestern flooding and Hurricane Katrina have raised concerns about reducing human and economic losses from flooding. In the United States, local governments are responsible for land use and zoning decisions that shape floodplain and coastal development; however, state and federal governments also influence community and individual decisions on managing flood risk. The federal government constructs some of the nation's flood control infrastructure, supports hazard mitigation, offers flood insurance, and provides emergency response and disaster aid for significant floods. In addition to constructing flood damage reduction infrastructure, state and local entities operate and maintain most of the flood control infrastructure and have initial flood-fighting responsibilities.

Chapter 4- Historically, floods have caused more economic loss to the nation than any other form of natural disaster. In 1968, Congress created the National Flood Insurance Program (NFIP) in response to rising flood losses and escalating costs resulting from ad-hoc appropriations for disaster relief. Federal flood insurance was designed to provide an alternative to federal disaster relief outlays by reducing the rising federal costs through premium collection and mitigation activities. The purchase of flood insurance was considered to be an economically efficient way to indemnify property owners for flood losses and internalize the risk of locating investments in the floodplains.

Chapter 5- Unusually cool, wet spring weather followed by widespread June flooding across much of the Corn Belt cast considerable uncertainty over 2008 U.S. corn and soybean production prospects. As much as 5 million acres of crop production were initially thought to be either lost entirely or subject to significant yield reductions. Estimates of flood-related crop damage varied widely due, in part, to a lack of reliable information about the extent of plant recovery or replanting in the flooded areas. These circumstances generated considerable market angst and U.S. agricultural prices for corn and soybeans, as reported on the major commodity exchanges, hit record highs in late June and early July. Since then, most of the Corn Belt has experienced nearly ideal growing conditions suggesting the potential for substantial crop recovery, and market prices have weakened accordingly.

In: Federal Flood Policy
Editor: James E. Rysanek

ISBN: 978-1-61324-017-5
© 2011 Nova Science Publishers, Inc.

Chapter 1

FEDERAL FLOOD POLICY CHALLENGES: LESSONS FROM THE 2008 MIDWEST FLOOD[*]

Nicole T. Carter

SUMMARY

Floods remain a significant hazard in the United States. Developing and investing in flood-prone areas represents a tradeoff between the location's economic and other benefits and the exposure to a flood hazard. In the United States, flood mitigation, protection, emergency response, and recovery roles and responsibilities are shared. Local governments are responsible for land use and zoning decisions that shape floodplain and coastal development. State and federal programs, policies, and investments influence community and individual decisions on managing flood risk. The federal government constructs some of the nation's dams and levees, offers flood insurance, supports nonstructural risk reduction actions (known as hazard mitigation), and provides emergency response and disaster aid.

In June 2008, a series of storms in several midwestern states caused $9 billion in damages. The 2008 flooding drew comparisons to the devastating 1993 Midwest flood and raised questions about whether the lessons from the 1993 flood were heeded. In 1993, hundreds of levees throughout much of the basin were breached in the Midwest causing $30 billion in damages; much of the damage was agricultural and occurred in soaked upland areas. In contrast,

---

[*] This is an edited, reformatted and augmented version of a Congressional Research Services publication, dated January 27, 2010.

the majority of the 2008 damages were concentrated along a few Mississippi River tributaries and in population centers with breached levees. The magnitude of the two floods simply overwhelmed the region's levees and dams, illustrating that some residual risk remains to people and investment behind these protective structures. Since 1993, emergency response and hazard mitigation programs have reduced risks in some Midwest communities; however, the region's flood risk continues to increase as more investments and people are located in flood-prone areas.

Since 1993, Congress, federal agencies, state, and local governments have taken steps aimed at reducing the nation's flood risk; at the same time, climate, population, and investment trends have increased the threat, vulnerability, and consequences of flooding. For example, Congress authorized using federal disaster assistance to cover more of the costs to acquire, relocate or elevate flood-prone homes and businesses. However, broader efforts to adopt a comprehensive flood policy and management strategy have not been pursued. The fundamental direction and approach of the national policies and programs remain largely unchanged since 1993. A comprehensive strategy would require regulation of floodplain use, significant changes to federal programs, and increased investment in flood risk reduction by all levels of government. Although they would reduce flood risk, these changes face significant opposition.

The 2008 Midwest flooding, Hurricane Ike in 2008, and Hurricane Katrina in 2005 have renewed interest in the suite of tools available to improve flood resiliency. The issue for Congress is deciding on whether and how to enact and implement feasible and affordable flood policies and programs to reduce flood risk. The challenge is how to structure federal actions and programs so they provide incentives to reduce flood risk without unduly infringing on private property rights or usurping local decision making. Tackling this challenge would require adjustments in the flood insurance program, disaster aid policies and practices, and programs for structural and nonstructural flood risk reduction measures and actions.

U.S. FLOOD CHALLENGE: A FEDERAL PRIMER

In late May and early June 2008, several midwestern states were hit with a series of storms that produced flooding along many Mississippi River tributaries and nearby segments of the Mississippi River. This flooding raised concerns about both the risk of another disaster like the devastating 1993 Midwest flood and the state of the nation's flood policies, programs, and infrastructure. Although emergency response has improved since 1993 and hazard mitigation programs have reduced some risks, the region's flood risk continues to increase as more investments and people are concentrated in flood-prone areas affected by extreme precipitation.

Riverine and coastal flooding remain serious risks to the nation's population and economy. The principal causes of floods in eastern states and the Gulf Coast are hurricanes and storms. Coastal counties are 17% of the land area, and home to roughly 50% of the country's population and jobs. Flooding in the Midwest and western states is primarily from snowmelt and rainstorms. At least 9 million homes and $390 billion in property are at risk from a flood with a 1% annual probability of occurring.[1]

Increasing flood hazards are putting existing developments at risk.[2] New development is occurring in flood-prone areas, often behind aging levees constructed to reduce agricultural damages rather than protect urban populations. National flood damages, which averaged $3.9 billion annually in the 1980s, nearly doubled in the decade 1995 through 2004. Total disaster assistance for emergency flood response operations, and subsequent long-term recovery efforts, increased from an average of $444 million during the 1980s to $3.75 billion from 1995 to 2004.

Congress and federal agencies have taken steps to address selected flood challenges; at the same time, climate, population, and investment trends have increased the threat, vulnerability, and consequences of flooding. In response to the 1993 flood, Congress shifted federal programs to increase support for a wider range of activities that reduce damage and prevent loss of life, such as moving flood-prone structures and developing evacuation plans; this broader set of activities is known as hazard mitigation. This shift has prompted wider use of nonstructural mitigation, particularly for new development and repairing damaged property. Traditional structural approaches, such as levees, floodwalls, and dams, continue to dominate much of the national investment in flood damage reduction. Often structural measures are the most readily available and locally acceptable tools to reduce flood risk for existing population, economic, and infrastructure hubs.

Since 2005, Congress has considered legislation and enacted other measures to address some flood issues; broader efforts to adopt a comprehensive flood policy and management strategy, however, have not been pursued. Hurricane Katrina's devastation in 2005 and the 2008 Midwest flood have again prompted attention to the suite of tools available to create a more flood-resilient nation. Many of these tools would require action by local governments, regulation of floodplain use, significant changes to federal programs, and substantially increased investment in flood damage reduction. Achieving these actions and implementing improved floodplain management is likely to confront opposition from those benefitting from the status quo and those opposed to land use regulation. And it likely would require broader congressional action than the incremental policy alterations that have been typical following recent floods.

This chapter first provides a primer on recent developments, the federal role in flood policy, and the limitations of levees and dams. The report then discusses lessons from the 2008 Midwest flood and contrasts the 2008 flood with the 1993 flood. It then discusses the evolution of U.S. flood policy, with particular attention to the role of Congress and federal agencies and programs, and the available tools for addressing the nation's flood challenge.

Recent Interest and Developments

The 2008 Midwest flood and the extensive damage and loss of life caused by Hurricane Katrina have raised awareness of flood risk, and levee construction and maintenance in particular. These disasters raised many flood policy questions, including whether to change the division of the roles and responsibilities between the federal, state, and local government; whether to have more federal leadership on floodplain management; and whether to increase coordination of federal flood-related actions.

Since Hurricane Katrina, Congress has conducted hearings and considered legislation on numerous aspects of federal flood programs and policies (see **Appendix A** for a list of flood-focused hearings since 1993). In the Water Resources Development Act (WRDA) of 2007 (P.L. 110-114), Congress enacted flood policy provisions aimed largely at improving the planning and safety of levees. Few other changes have been enacted, and the legislation considered has largely addressed individual programs or agencies, rather than attempting a comprehensive realignment of federal flood actions.

Two recent developments may garner congressional attention. In January 2009, the National Committee on Levee Safety (created by WRDA 2007) released its draft recommendations for a national levee safety program. On January 15, 2009, Congress received a report on the Upper Mississippi River Comprehensive Plan (UMRCP) study; the report identifies the costs and benefits of significantly increasing the level of flood damage reduction along the mainstems of the Mississippi and Illinois Rivers. Both developments are discussed later in this CRS report.

Flood Policy in a Federalist System: Shared Responsibilities

In the United States, flood-related roles and responsibilities are shared; local governments are responsible for land use and zoning decisions that shape floodplain and coastal development, but state and federal governments also influence community and individual decisions on managing flood risk. State and local governments largely are responsible for making decisions (e.g., zoning decisions) that allow or prohibit development in flood prone areas. Local and some state entities construct, operate, and maintain most levees and have initial flood-fighting responsibilities. Levees are embankments built alongside a river to prevent high water from flooding bordering land.[3] The federal government constructs some of the nation's levees and dams in partnership with local project sponsors, but turns over operation and maintenance responsibility for most of these levees to local entities. The federal government also supports hazard mitigation, offers flood insurance, and provides emergency response and disaster aid for significant floods.

Federal flood programs and investments consist primarily of:

- Construction investments in select dams, levees, seawalls, and beach improvements;
- Nonstructural hazard mitigation assistance;
- Flood and crop insurance; and
- Disaster preparedness, response, and recovery assistance.

The principal federal agency involved in levee construction and repair is the U.S. Army Corps of Engineers (Corps). (See Appendix A for a table of selected congressional direction to guide the Corps' efforts in flood damage reduction.) Other federal agencies also are involved with flood-related activities, such as the U.S. Department of Agriculture's Natural Resources Conservation Service, the Department of the Interior's Bureau of Reclamation, and the Tennessee Valley Authority. The Federal Emergency Management Agency (FEMA) has primary responsibilities for federal hazard mitigation, flood insurance, and disaster assistance. FEMA and the Corps require levee inspection and certification for participation in the Corps' Rehabilitation and Inspection Program (RIP, also known as the P.L. 84-99 program, which is discussed on page 11) and FEMA's National Flood Insurance Program (NFIP). Crop insurance is administered by the U.S. Department of Agriculture.

Limits to Levees and Dams

Hurricanes, other severe weather systems, and rapid snowmelt can cause flooding. Floods are a vital element of variability in the hydraulic regime of healthy riverine, estuarine, and coastal ecosystems; however, they can result in immediate human suffering and economic loss. Failure of levees and dams and inadequate urban drainage also may result in flooding.

Hurricane Katrina focused attention on the performance of levees and floodwalls and the risk remaining behind these structures. There are over 100,000 miles of levees in the nation, only 14,000 miles of these receive regular inspections by the Corps. These levees do not work in isolation from the rest of the watershed. Levees restrict the size of the floodplain which constricts floodwaters to a smaller area, thus raising river crests and often increasing the river's velocity. How land is used can have a dramatic impact on the response of streams to flooding (e.g., tile drains in agricultural areas, impervious areas in urban developments can increase runoff and flood crests). Land use choices can cause 500-year flood levels to be produced by events of lesser magnitude. Some land uses can, therefore, result in levees having to hold back higher flows more frequently.

Generally the Corps no longer refers to levees and dams as "flood control" measures, rather it calls levee projects "flood damage reduction" measures and discusses them in the context of a suite of "flood risk management" actions. This language shift reflects an appreciation of the limitations of these structures. Levees, if constructed properly, should perform up to their design level of protection (i.e., 100-year level of protection is the design to reduce damages from a flood with a 1% probability of occurring in a given year); however, when a flood is greater than that design, the levees are overtopped. Sufficient overtopping often results in levee failure, known as breaching. Similarly, dam are designed to spill floodwaters when their capacity is exceeded.

Although floodwaters overtopped and breached many Midwest levees and a few dams in 1993 and 2008 causing significant economic damage, the dams and levees worked largely as designed.[4] The dams reduced the river crests, and many levees held, thereby preventing floodwaters from damaging many population centers and agricultural and industrial investments. Nonetheless, the potential role of the basin's levees in increasing damages because of their encouragement of risky development and reduction natural flood storage remains debated and part of the active discussion about the future of the basin's floodplains.[5]

The performance of the Midwest levees contrasts to the performance of floodwalls in New Orleans during Hurricane Katrina. Some of the floodwalls protecting urban New Orleans failed before their design level was reached, and the damage was catastrophic. These floodwalls lost their integrity, allowing the water level in the city to rise to the level of surrounding water bodies.

Residual risk is the portion of risk that remains after flood damage reduction structures have been built and other damage-reducing measures have been taken. Risk remains because of the likelihood that levees and dams will be overwhelmed by severe floods and the risk of structural failure. The damaging consequences of floods increase as development occurs behind levees and below dams; ironically, this development may occur because of the flood protection provided. The nation's risk in terms of lives lost, economic disruption, and property damage is increased by overconfidence in the level and reliability of structural flood protection.

The inability of infrastructure to protect against all flooding is fundamental to understanding why some flood risk always remains and to making decisions of how to prioritize flood risk reduction investments. Decision-makers are faced with choosing the level of protection to provide for urban areas, critical infrastructure, rural areas, etc., and making tradeoffs when distributing limited funds across different projects throughout the nation and across the range of flood damage reduction measures (e.g., levees, buyouts, insurance).

2008 MIDWEST FLOOD: WHAT HAPPENED AND HOW DOES IT COMPARE TO 1993?

Intense Precipitation in Tributary Watersheds in June 2008

Intense precipitation in May and early June 2008 led to numerous record and near-record river crests in the Midwest, particularly on Mississippi River tributaries in eastern Iowa and southern Wisconsin. The resultant flooding was localized, but extremely severe.[6] A few streams, particularly in eastern Iowa, had discharges that exceeded record levels for ten or more consecutive days. The 100-year and 500-year flood levels were exceeded in much of eastern Iowa. These exceptional flood levels overtopped levees and flooded areas that many people assumed to be safe.

The most affected tributaries were the Cedar, Des Moines, and Iowa Rivers. Record river stages were set at 47 river gage stations on more than 12 tributary rivers and creeks. In some locations, the new record crests were considerably higher than the previous record crests, including 1993 records. Levees in Des Moines and Cedar Rapids were breached. Two significant examples were the Cedar River at Cedar Rapids (see Box 1) and the Iowa River at Columbus Junction and Iowa City.[7] As the floodwaters from these tributaries entered the mainstem of the Mississippi River, they set records at Keithsburg and Gladstone, Illinois and Burlington, Iowa, and approached record stages at other locations.

Box 1. Cedar River Overwhelms Cedar Rapids Levees Causing Extensive Damage

The damage to Cedar Rapids by the 2008 flood was extensive. The river crest rose to more than 31 feet, well above the estimated 500-year flood level and 12 feet above the 1993 crest. The floodwaters easily overwhelmed the city's levees which stood at 22 feet. This crest exceeded the previous record set in the 1850s, when the river reached 20 feet. The flood inundated 9.2 square miles; 1,300 city blocks; 3,894 single family residences; and 818 commercial properties and government buildings. Because the floodwaters reached locations far outside the 100-year floodplain, many homes not covered by NFIP policies were inundated. Up to 400,000 cubic yards of trash and debris were expected to be generated during clean-up, more than the city produces in an entire year.

rce: CRS adapted from Midwest Regional Climate Center, *Midwest Weekly Highlights - June 17-23, 2008*, available at http://mcc.sws.uiuc.edu/cliwatch/ 0806/ 080623.htm.

Storms in 2008 Were Quick, Which Caused Primarily Tributary Flooding; Extended Storms in 1993 Inundated the Region

The 1993 flood is sometimes described as a "leisurely" disaster because it resulted not from a single storm but from a weather pattern that remained stationary for months.[8] From May through September of 1993, record or major flooding occurred across North Dakota, South Dakota, Nebraska, Kansas, Minnesota, Iowa, Missouri, Wisconsin, and Illinois. The geographic scale of the flood was vast, much larger than the 2008 flood. The four-month duration of the 1993 flooding significantly increased the scale of its consequences.

In 1993, extensive reaches of the mainstems of the Missouri and Mississippi Rivers experienced flooding of extended duration (see Figure 1). Lower reaches of the Illinois River and extended reaches of the Kansas, Des Moines, and Iowa Rivers also recorded record flood crests. Approximately 600 river forecast points in the Midwest were above flood stage at the same time. In contrast, the 2008 flooding was shorter and concentrated along select Mississippi River tributaries and nearby segments of the Mississippi River.

The 1993 differs from the 2008 flood in its areal extent, magnitude, duration, volume of floodwater, extent of damage, and time of the year. The rainfall causing the 1993 flood was uncommonly persistent and covered a huge drainage area encompassing most of nine states. This scenario caused many tributaries to crest at about the same time and to synchronize with crests on the mainstem of the Mississippi and Missouri Rivers.

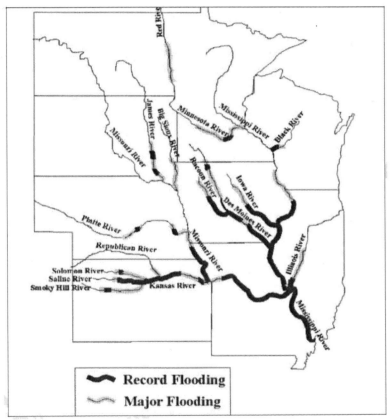

Source: Interagency Floodplain Management Review Committee, *Sharing the Chalenge: Floodplain Management into the 21st Century* (July 1994), available at http ://www.floods.

Figure 1. 1993 Midwest Flood: Major and Record Flooding on the Mississippi River and Its Tributaries.

2008 FLOOD DAMAGES WERE CONCENTRATED IN DURATION AND EXTENT; 1993 DAMAGES CREATED A REGIONAL ECONOMIC DISASTER

Forty-eight deaths and economic damages of $30.2 billion were attributed to the 1993 flood;[9] more than 70,000 homes were damaged.[10] Roughly150 major rivers and tributaries had flooded, at least 15 million acres of farmland had been inundated. More than half of the economic losses were agricultural.[11] It is important to note that most agricultural damage resulted from wet fields in upland areas and a short growing season, rather than inundation by floodwaters.

According to a 1995 analysis of Corps records by the then-General Accounting Office (renamed the Government Accountability Office, GAO), many mainstem levees withstood the 1993 floodwaters, preventing both flooding of an additional 1 million acres and an additional $19 billion in damages.[12] Other levees in the basin were overtopped when floodwaters exceeded their design. Four levees that were regularly inspected by the Corps were breached or otherwise allowed water into protected areas before their design capacity was exceeded.

In 1993, transportation impacts were severe and lengthy. Barge traffic on the Missouri and Mississippi Rivers was stopped for nearly 2 months. Bridges were out or not accessible on the Mississippi River from Davenport, Iowa, downstream to St. Louis, Missouri. On the Missouri River, bridges were out from Kansas City, Missouri, downstream to St. Charles, Missouri. Major east-west rail and road transportation routers were severed, causing significant delays and rerouting. Numerous interstate highways and other roads were closed. Ten commercial airports flooded. Much of the railroad traffic in the Midwest was halted. Other public infrastructure, such as sewage treatment and water treatment plants, was damaged or destroyed.

The 2008 floodwaters caused local disaster conditions and significant damages. Property, agricultural, and other damages are estimated at $15.0 billion, and the weather was attributed to 24 deaths.[13] Unlike in 1993, damage in 2008 was from inundation by floodwaters along the rivers, not in saturated upland areas.

The magnitude and severity of the 1993 flood event was overwhelming. Hundreds of levees were breached along the Mississippi and Missouri Rivers; in contrast, dozens of levees were breached in 2008. The levees breached on the Mississippi mainstem in 2008 were primarily lower agricultural levees. In 2008, although some levees overtopped, they worked largely as intended; that is, they held back floodwaters until the floodwaters exceeded the level of protection the levees were expected to provide. Many of these overtopped levees were protecting primarily agricultural areas and provided the anticipated 5- to 25-year protection. Floodwaters overtopping levees protecting larger communities, like Cedar Rapids, resulted in considerable and concentrated damage; these damages contributed to the 2008 damage estimates being half of the 1993 flood damages even though the duration and extent of flooding was less than in 1993.

The lower regional damage estimates in 2008 ($15 billion compared to $30 billion in 1993) fail to capture the challenge of recovery in severely affected communities. The social and economic consequences for families and communities can be extreme, and recovery in severely damaged communities often takes years.

Some roads in eastern Iowa and northwest Illinois sustained severe flood damage in 2008, resulting in closings, delays, and lengthy detours.[14] Major rail lines in Iowa, Wisconsin, Minnesota, Missouri and Illinois were washed out. Navigation locks 13-25 on the Mississippi River closed, leaving 281 miles of the river closed to barge traffic. In June, the flooding was predicted to have major impacts on agriculture. CRS Report RL34583, *Midwest Floods of 2008: Potential Impact on Agriculture*, found that anticipated crop losses contributed to agricultural prices for corn and soybeans hitting record highs in late June and early July. After that, however, most of the "Corn Belt" experienced nearly ideal growing conditions resulting in substantial crop recovery and lower market prices. Therefore, although the floodwaters caused transportation and agricultural disruptions, they largely were resolved and repairs were underway once the localized flooding diminished.

LESSONS FROM THE 2008 MIDWEST FLOOD

Post-1993 Investments Paid Off, but More Development at Risk

Is the Midwest more or less at risk of floods now than in 1993? Some communities in the Midwest are less at risk than in 1993 due to buyouts, relocation, and floodproofing (i.e., adjustments to structures that reduce or eliminate flood damage) of vulnerable properties. Relocation of key public infrastructure such as drinking water facilities reduced the consequences of flooding. The general sense is that flood risk reduction in the Mississippi River basin since 1993 paid off in 2008.[15] However, the basin's aggregate flood risk appears to be increasing.

After the 1993 flood, the GAO found that not only had man-made changes within the basin over many decades raised the levels of floodwaters in the basin's rivers, but also the precipitation trend in the basin appeared to be increasing over the long term.[16] Congress reacted to the 1993 flood by enacting a number of policy and program changes. It authorized using a portion of federal disaster assistance to cover 75% of the cost to acquire, relocate or elevate flood-prone homes and businesses; prior to the change, the federal cost share had been 50%. Buyouts of at-risk properties using FEMA disaster mitigation funds were more extensive for the 1993 flood than for previous disasters. In the nine states that flooded, FEMA ultimately moved more than 300 homes, and bought and demolished nearly 12,000 at a cost of over $150 million. The lands were turned to flood-friendlier uses such as parks and wildlife habitat. State and federal agencies have also acquired interest in over 250,000 acres of flood-prone land, principally frequently flooded farmland. Another 9,140 properties in 140 communities were elevated, acquired or relocated under hazard mitigation grants. Taken together, these actions signaled a dramatic departure from historic flood policies, which relied primarily on levees and dams.

Significant redevelopment and new development has occurred in Midwest floodplains since 1993, including areas flooded in 1993. The population in the 500-year floodplain has increased by 17%; the population in the area flooded in 1993 has increased by 18%.[17] There also has been significant new commercial and industrial development and highway and interchange development within the 500-year floodplain.[18] New development in the 100-year floodplain would be required to meet floodplain development regulations if the community participates in the NFIP. Therefore, this development ought to be more flood resilient than before. Development or redevelopment, however, may not be more flood resilient if built behind levees that are certified to provide 100- year protection or if the community does not participate in the NFIP. Therefore, risk from the 100-year flood in the NFIP-participating communities in the Midwest may have decreased since 1993, but the risk in these communities to more intense floods may have increased. Moreover, risk in the 500-year floodplain has increased substantially due to development.

Better Forecasting Data Needed to Improve Emergency Response

Rainfall and streamflow data are fundamental to coordinating and managing emergency flood response activities. In 2008, several Mississippi River tributaries rose quickly. At the most severely affected locations, rivers rose at a rate of one foot per hour. River crests on some tributaries eventually exceeded their river gauges ability to record. Limited river gauging information constrained the National Weather Service and others in developing timely and accurate river stage forecasts.

In October 2008, the Corps convened a Rainfall-River Forecast Summit of representatives of the Corps, the National Weather Service, and the USGS; the summit also included a public meeting.[19] Federal summit participants concluded that significantly more rainfall fell than was predicted resulting in record river flood stages that were not forecast with sufficient lead time for appropriate emergency response preparations. Although the coordination and data exchange generally went well, according to summit participants, discrepancies of reported data created forecasting challenges and raised doubts of forecast reliability. River gauges swept away by the floodwaters resulted in data gaps during critical periods. As a result, some river forecasts were inaccurate. Better coordination, communication, and collaboration, as well as more and better data measurements, were recommended by the summit participants.

Gains in Managing Levee Repairs, but Levee Deficiencies and Improvements Remain Challenges

During the 2008 flood, a total of 41 levees overtopped. Of these, only six were constructed by the Corps; these had been turned over to a local entity for maintenance. Another 19 of the 41 were constructed by local entities but met participation requirements for the Corps' Rehabilitation and Inspection Program, which assists with repairs.[20] The other 16 overtopped levees were built and operated by local entities and had not met RIP participation requirements.[21] Their repair is not eligible for federal assistance through RIP, but may be eligible for some other federal assistance through the Natural Resources Conservation Service for levees in small watersheds or FEMA, particularly if there is an immediate threat to life and property.

Repairing levees following the 2008 floods has illustrated some improvements since 1993, as well as continuing issues with repair and maintenance. Reducing flood risk to conditions prior to the damaging event can be complicated by choices about whether and how to repair damaged levees and the availability of assistance from various federal agencies. At the same time these choices represent opportunities to consider alternative methods of managing flood risk.

In 2008 coordination of near-term alternatives for levee repair showed improvement over 1993. The Corps is leading a regional Interagency Levee Task Force for the 2008 Midwest flood. This type of task force has not been used extensively before. The basis for its use is a February 1997 guidance memo issued by the Office of Management and Budget and the Council on Environmental Quality. The memo was published as part of the Clinton Administration's efforts to improve flood and floodplain management policy after the 1993 floods. The memo instructed federal agencies to "fully consider relevant options, including non-structural alternatives, during evaluation and review of levee repair and reconstruction projects...".[22]

The 2008 Midwest Interagency Levee Task Force was established to assist in the rapid and effective recovery of floodplain management systems in the affected communities and areas before the next flood season. The task force is charged with implementing a collaborative and integrated regional approach by the federal agencies to the long-term restoration of damaged floodplain management systems. Its use is viewed by floodplain management advocates as promising because it is not only looking at rebuilding levees but also considering long-term mitigation and recovery.

A common issue that arises following a flood is local interest in not only repairing levees but improving the level of protection provided. Rehabilitation and Inspection Program funds are expressly restricted to repairing and cannot be used to increase protection. The RIP program is not designed to evaluate the federal interest in investments to further reduce the local flood risk. If federal participation is sought in increasing protection, the traditional process is to initiate a Corps flood damage reduction study. This is separate from repair work.

Interest in increasing the resiliency of levees and their level of protection has become more salient since the 2005 hurricane season. Since 2005, the levee inspection and certification programs used by the Corps for the Rehabilitation and Inspection Program and FEMA for the NFIP have been strengthened to address weaknesses identified in the programs. Consequently, significant numbers of levees have been identified as deficient since 2005. If the deficiencies are not addressed, the levee may not be eligible for federal repair assistance if damaged by a flood, and NFIP floodplain requirements may go into effect (e.g., areas behind the levee may be required to purchase flood insurance). The local entities that own and maintain the levees are responsible for making the improvements necessary to pass inspection and obtain certification. Generally federal funds are not available for these rehabilitations which are considered part of the local responsibility for levee upkeep. Additionally NFIP hazard map modernization and policy changes have improved the understanding of current risks, resulting in some areas receiving higher risk designations and having stricter NFIP requirements apply.[23]

No Comprehensive Midwest Flood Management Strategy in Place

The dams and levees of the Upper Mississippi River System were largely constructed as separate projects, not in accordance with a basin flood damage reduction plan. The existing facilities have varying structural integrity, and provide varying levels of flood risk reduction for similar land uses.[24] The levels of protection range from less than 5 years up to 500 years, with three-quarters of the urban systems designed to manage a 500-year flood. Land use and flood management changes (e.g., levee building that constricts the flow of floodwaters to within the levee banks, or channel straightening that increases the velocity of floodwaters) in upstream areas can alter flood risk in downstream areas. Whether and how to integrate Midwest flood management and related infrastructure was an issue after the 1993 flood. Nonetheless, responsibilities for flood programs in the basin remains largely unchanged (i.e., distributed among local, state, and federal entities).

Like many other basins, no broad planning authority has guided the Upper Mississippi basin's water resource management since the termination of the Upper Mississippi River Basin Commission (UMRBC) at the end of 1981, which had been established in 1972.[25] The UMRBC was a regional entity for comprehensive planning that integrated federal-state-local planning with public input. The UMRBC prepared a comprehensive master plan for management of the upper Mississippi River system's water and related-land resources. The Commission's termination complicated implementation of the master plan. The interstate Upper Mississippi River Basin Association (UMRBA), which was founded in 1981 remains in operation; its role largely has been limited to a policy research and coordination forum for the basin states. Because the UMRBA is a state initiative, unlike the UMRBC, the federal government has no voice.

The long-standing Mississippi River Commission has authority for river improvements from the Mississippi River's delta to the headwaters.[26] The Commission provides water resources engineering direction and policy advice to the Administration, Congress, and the Army by overseeing the planning and reporting on river improvements. Unlike in the lower basin, the Commission currently does not have the funded authority to implement improvement plans in the upper Mississippi River. In 1997, the Commission initiated a process of listening, inspecting, and partnering in the upper basin, but has not pursued significant steps to increase its upper basin role.

UPPER MISSISSIPPI FLOOD MANAGEMENT

Post-1993 Flood Proposals and Recommendations

The 1993 flooding engendered some congressional interest in a systemic approach to flood damage reduction on the upper basin. Following the 1993 flood, numerous reports were produced recommending changes to various aspects of how floods are managed in the United States and the Midwest in particular. The most prominent of these reports was the July 1994, *Sharing the Challenge: Floodplain Management into the 21st Century*, by the Interagency Floodplain Management Review Committee, often called the Galloway report after the Committee's chairman, Brigadier General Gerald Galloway.[27] Box 2 briefly describes the report's recommendations for the Upper Mississippi River and a general floodplain management strategy.

In August 1994, S. 2418 (103rd Congress) was introduced. It would have acted on many of the report's recommendations. If enacted, it would have represented a significant shift in flood and floodplain management for the Midwest. The legislation would have required development of comprehensive river basin management plans for the long-term ecological, economic, and flood management needs of the Upper Mississippi and the Missouri Rivers and established federal-state coordinating committees to review and recommend the basin plans. The bill also included numerous other broad water resources policy provisions that would have emphasized nonstructural measures for risk reduction. This legislation was not enacted.

Upper Mississippi River Comprehensive Plan

It was not until the Water Resources Development Act of 1999 (P.L. 106-53) that a new flood management study for the upper Mississippi River basin was authorized. In Sec. 459 of WRDA 1999, Congress authorized the Upper Mississippi River Comprehensive Plan (UMRCP). It directed the Secretary of the Army to "develop a plan to address water resource and related land resource problems and opportunities in the upper Mississippi and Illinois River basins, from Cairo, Illinois, to the headwaters of the Mississippi River, in the interest of systemic flood damage reduction....". The Corps chose not to perform a comprehensive watershed analysis encompassing the entire upper Mississippi River basin and its tributaries, citing fiscal and time constraints. Instead, it limited the study to the mainstem Mississippi River and Illinois River floodplain. The Missouri River and smaller tributaries, such as the Cedar River and Iowa River, were excluded. This scope left out the majority of the areas most severely affected in 2008.

Box 2. Recommendations for a Comprehensive Upper Mississippi Strategy from a Report on the 1993 Flood

The 1994 Galloway report recommended a floodplain management strategy that sequentially supported:
- avoiding inappropriate use of the floodplain,
- minimizing vulnerability to damage through both structural and nonstructural means, and
- mitigating flood damages when they do occur.

It also included a recommendation to reduce the vulnerability of population centers to roughly the 500-year flood. For the Upper Mississippi basin specifically, the report's recommendations included:
- establishing upper Mississippi River and Missouri River basin commissions with a charge to coordinate development and maintenance of comprehensive water resources management plans to include, among other purposes, ecosystem management, flood damage reduction, and navigation; and
- expanding the mission of the Mississippi River Commission to include the Upper Mississippi and Missouri rivers (to recognize ecosystem management as a co-equal federal interest with flood damage reduction and navigation, commission membership should be expanded to include the Department of the Interior); and
- assigning responsibility for development of an Upper Mississippi River and Tributaries system plan and for a major maintenance and major rehabilitation program for federally related levees (including those participating in RIP) to an expanded Mississippi River Commission, operating under the Corps.

The UMRCP final report, dated June 2008, was transmitted to Congress on January 15, 2009.[28] Congress must decide how to proceed given the analysis presented in the report. The UMRCP was conducted as a preliminary study, similar to the level of detail in a Corps reconnaissance study. The UMRCP final report and supporting documents are not at the level of detail of a feasibility study, which typically informs decision-making on construction authorization.

The report states "additional authority to implement the Comprehensive Plan is not being recommended nor requested at this time based upon the [national economic development] evaluation of alternative plans." Although the report does not recommend proceeding with additional authority to implement the comprehensive plan, the report identified the Corps' preferred alternative; it would provide 500-year protection at a total cost $4.42 billion.

The Assistant Secretary of the Army (Civil Works) in his January 2009 transmittal letter to Congress also did not recommend proceeding with implementation; the letter stated that "recommendations for implementation of a specific plan based on a reconnaissance level of detail is premature."[29] Instead the Assistant Secretary recommended some intermediate steps — expanding the UMRCP to include Mississippi River tributaries, conducting cost-shared studies of the reconstruction needs for the basin's existing flood damage infrastructure (where a federal interest is identified), and conducting a study of flood protection for critical transportation infrastructure such as bridge approaches and railroads.

Earlier in August 2008, the Mississippi River Commission in its planning oversight and policy advice role voted to support implementing the preferred alternative.[30] The Commission believed that the full benefits of implementing the preferred alternative were not adequately measured with the current Corps project planning guidelines.[31] Also in 2008, the then-Governor of Illinois and the then-Governor of Missouri wrote letters of support for a comprehensive plan; these letters, however, supported an alternative that was studied but not the preferred alternative.

Congress is faced with deciding how to proceed given the differing recommendations of the Corps report and the Assistant Secretary of the Army, Mississippi River Commission, two state governors, and the many stakeholder viewpoints in the basin. For example, some stakeholders prefer emphasizing nonstructural measures to manage flood hazards, and others are concerned about tributary flood risk. Appendix B provides an analysis of the UMRCP final report and its limitation, the preferred alternative identified in the report, various stakeholders recommendations on how to proceed, the debate over the future of the Midwest flood and floodplain management, and the potential state and federal roles.

FEDERAL FLOOD POLICY SINCE 1993: TOUGH CHALLENGES REMAIN

Unfinished Business on Many Post-1993 Recommendations

Although Congress did take numerous actions after the 1993 flood to improve flood policy and programs, numerous recommendations in the Galloway report have not been implemented, including:

- Enact a national Floodplain Management Act to define government responsibilities, strengthen federal-state coordination and improve accountability. It should establish a national model for floodplain management that recognizes the states as the principal floodplain managers.
- Reactivate the Water Resources Council to coordinate federal-state-tribal water resources activities.
- Reestablish a river basin commission, as needed, as forum for coordination of regional issues.
- Issue a new Executive Order to reaffirm the federal government's commitment to floodplain management with a broader scope and more defined federal agency responsibilities than in the existing floodplain E.O. 11988.
- Limit public grant assistance available to communities not participating in the NFIP.
- Provide loans for the upgrade of infrastructure and public facilities.
- Reduce the vulnerability of population centers to damages from roughly the standard project flood (which is roughly the 500-year flood).

Many of the actions that were taken were among the narrower recommendations of the Galloway report, such as increasing the waiting period for flood insurance policies to become active.

Flood and Floodplain Management Policy

Over the years, many commissions and reports, like the Galloway report, have called for a fundamental reorientation in national flood policy that addresses not only the economic but also the social and environmental welfare tradeoffs of floodplain development.[32] These commissionsand reports have urged Congress, relevant agencies, and the public to commit to the broad goal of reducing the dangers and damages via flood and floodplain management, rather than allowing development that could be located elsewhere to occur in flood-prone areas. Despite these recommendations, a fundamental reorientation for floodplain management has not occurred.

Although federal efforts have not been guided by a clearly defined flood policy or floodplain vision, many incremental changes to improve flood programs and projects have been enacted or adopted at all levels of government. These actions include supporting nonstructural flood damage reduction, retiring flood-prone farmland, purchasing repetitive flood loss properties, augmenting hazard mitigation activities, fostering floodplain regulation, and guiding federal actions in floodplains.

Notwithstanding these changes, the nation's flood risk is increasing. Many of these changes have seen only marginal implementation, enforcement, and funding. The incremental improvements largely have been overwhelmed by incentives to develop floodplains and coastal areas and by a growing population, or have never fully implemented or enforced. Other federal actions produce some indirect flood risk reduction benefits; for example, Congress has supported conservation efforts on agricultural lands and wetlands protection that may reduce flood damages by slowing down or temporarily storing flood waters. Whether these benefits are overwhelmed by changes in flood-prone land use (e.g., conversion of agricultural land behind levees to residential development) remains largely unknown because regional-scale and multi-agency plans and evaluations have been rare.

Generally, congressional oversight, administrative implementation, and federal appropriations have reflected a reactive and fragmented approach to flooding. Earlier institutional arrangements that provided avenues for more coordinated federal flood efforts have diminished. For example, the national-level Water Resources Council which was established by the Water Resources Planning Act (P.L. 89-80), disbanded in 1983; the Federal Interagency Task Force on Floodplain Management, which had continued some of the Water Resources Council flood-related functions after 1983, stopped convening in the late 1990s. Federal support and opportunities for local capacity building decreased with the loss of these institutions.

Flood policy continues to be dominated by separate treatment of structural flood damage reduction investments (e.g., levee building), the NFIP, and federal disaster aid, rather than a comprehensive flood risk and floodplain management approach. Current arrangements of aid, insurance, and water resources projects at times unintentionally provide disincentives to reduce exposure to flood risks. This is in contrast to recommendations promoting a focused and coordinated effort to reduce the cost of flooding on the economy, improve public safety, and promote state and local capacity and responsibility for flood management.

Federal Flood Insurance and Mapping

In 1968, Congress created the National Flood Insurance Program as an alternative to disaster assistance and to manage the escalating cost of repairing flood damage to buildings and theircontents. Under the NFIP, the federal government identifies and maps areas subject to flooding, provides insurance to property owners in flood-prone areas, and offers incentives for communities to reduce future flood-related losses through floodplain management measures. Since 1973, homeowners in 100-year flood-prone areas are required to buy flood insurance if using a federally backed mortgage. Today, the NFIP provides flood insurance to more than 5 million homeowners, renters, and business owners in over 20,000 participating communities.

A significant policy reaction to the 1993 flood was passage of the National Flood Insurance Reform Act of 1994.[33] The flooding revealed that most flooded homeowners did not have flood insurance. And mortgage lenders had been lax in checking if federally backed mortgages were being granted in flood-prone areas, as required by NFIP. The 1994 legislation aimed to improve compliance with NFIP's mandatory flood insurance requirement, and to pressure lenders to ensure that at-risk owners in a flood zone purchase insurance. The legislation also:

- Created the Increased Cost of Compliance program within the NFIP. This program gives money to insured owners of substantially damaged properties to meet the more expensive costs of rebuilding according to a local floodplain management ordinance.
- Created the Flood Mitigation Assistance program. This program funding is derived from a surcharge added to all flood insurance policies nationwide. The funds are distributed as grants to states for flood mitigation.
- Increased emphasis on floodplain mapping.
- Codified the Community Rating System (CRS) into the NFIP. The CRS is an incentive program to reduce communities' flood insurance

premiums by exceeding the minimum flood risk reduction requirements of the NFIP.

After the 1993 floods, Congress authorized FEMA to use a portion of federal disaster assistance to cover 75% of the cost to acquire, relocate or elevate homes and businesses; set aside flood insurance premiums to relocate flood-prone buildings; and tighten flood insurance purchase requirements.[34] These actions signaled a shift toward hazard reduction away from reliance on levees and dams. Nonetheless, the potential consequences of floods are increasing as more people and investments are located in flood-prone areas.

Some of the more significant changes in flood-related policy have consisted of efforts to improve the NFIP (e.g., improvements to increase participation in the program and better manage repetitive loss properties) and reorganization of federal emergency response and recovery following the 9/11 attacks and Hurricane Katrina's impact on New Orleans. Considerable concerns continue to be raised about the degree of subsidization under the NFIP and the financial foundation of the program. Numerous GAO studies have reviewed various aspects of the NFIP; some recommendations have been implemented. In 2006, an independent review working group released its evaluation of the NFIP; the recommendations are among other changes that have been considered, but not enacted. Reorganization of emergency response, in particular the placement of FEMA within the Department of Homeland Security, remains a topic of much debate.

Flood Map Accuracy

As part of the NFIP, FEMA has implemented a standardized flood mapping program covering a large fraction of the population at risk. Government agencies use these maps to establish zoning and building standards and to support transportation, infrastructure, and emergency planning. Insurance companies, lenders, realtors, and property owners use maps to determine flood insurance needs and to assess their flood risk.

In January 2009, the National Research Council released *Mapping the Zone: Improving Flood Map Accuracy*. The report calls for investments in improving the accuracy of NFIP maps. It cites maps as central to anticipating, preparing for, and insuring against flooding. It found that current maps have significant uncertainties and do not necessarily represent current floodplain conditions. The Council concluded that these investments are needed and economically justified despite recent investments. From 2003 to 2008 at a cost of more than $1 billion, FEMA and local and state partners collected new flood data in unmapped areas, updated existing data, and digitized flood maps that were previously on paper. The Council found that although 92% of the continental U.S. population now has digital flood maps, only 21% has maps fully satisfying FEMA's data quality standards.

To remain accurate, flood maps must require updating to reflect changes in the flood threat (e.g., changes in sea level or precipitation patterns) and land use changes that affect flood risk. Future conditions (e.g., anticipated sea level rise, changes in hydrology due to land use changes) currently are not considered in developing NFIP maps.

Trends Affecting Flood Risk

Climate, Demographic, and Development Trends

Growth in total damage from floods in the United States since the early 1930s can be attributed to both climate factors and societal factors: that is, increased damage associated with increased precipitation and with growth in population and wealth.[35] Much of the flood-related damage in recent decades is the result of numerous human choices, meaning that society has considerable potential to reduce flood risk. Without major changes in societal responses to weather and climate extremes, it is reasonable to predict ever-increasing losses even without any detrimental climate changes.[36] As the former General Counsel of FEMA put it:

> The challenge is that more and more development is taking place in flood prone and hurricane prone areas. People like to live near the seashore. But unless the actual cost of living by the water is reflected in the cost of ownership — including the cost of building property to resist wind damage, elevating out of floodplains, and insuring at actuarial rates for the cost of rebuilding after inevitable floods and hurricanes — the result will only be more development in more risk prone areas ...[37]

Climate and population trends are combining in coastal areas so that flood risks of coastal storms exceed river flooding risks. The top eleven amounts paid for NFIP claims were for coastal storms (including Hurricane Ike). The 1993 Midwest flood ranks twelfth, and the 2008 Midwest flood is not in the top 20 NFIP events.

Coastal Vulnerability

Damage caused by Hurricane Katrina and other coastal storms illustrates the vulnerability of the nation's coastal developments to storm surge, flooding, erosion, and other hazards. Hurricane-prone states have increasingly dominated NFIP outlays and disaster losses. The risk facing the nation's coastal development, particularly barrier islands and other particularly vulnerable locations, is great regardless of whether climate change may alter the intensity and frequency of hurricanes. Severe storms and their surges have plagued coastal communities for centuries, costing thousands of lives, and damaging communities, businesses, and infrastructure.

Since the mid-1960s, the federal role in coastal hurricane storm protection has become more prominent; the Corps, with nonfederal sponsors, builds structures and places sand periodically for beach renourishment to reduce flooding. Congress also has enacted laws aimed at protecting coastal resources that have some flood risk reduction benefits. Through reauthorizations and amendments to the Coastal Zone Management Act of 1972 (P.L. 92-532) and the Coastal Barrier Resources Act of 1982 (P.L. 97-348), Congress has tried to improve federal actions that support coastal resource protection.[38] With the passage of the Coastal Zone Management Act in 1972, Congress was responding primarily to widespread public concern about estuarine and oceanfront degradation; the act provides for federal assistance to state and local coastal zone. The Coastal Barrier Resources Act prohibits federal spending that would support additional development in designated relatively-undeveloped coastal barriers and adjacent areas. Notwithstanding these efforts, both increasing coastal populations and the dominance of NFIP claims and federal disaster aid to coastal states indicate that significant coastal flood risk remains.

RECENT CONGRESSIONAL STEPS TO ADDRESS THE FLOOD CHALLENGE

The 2008 Midwest flood, Hurricane Katrina, and other levee breaches have increased the congressional debate about how to manage flood and infrastructure risks, what is an acceptable level of risk—especially for low-probability, high-consequence events—and who should bear the costs to reduce flood risk (particularly in the case of levee construction and rehabilitation). Issues to be addressed include protecting concentrated urban populations, reducing risk to the nation's public and private economic infrastructure, reducing vulnerability by investing in natural buffers, and equity in protection for low-income and minority populations. A challenge for Congress is structuring federal actions and programs so they provide incentives to reduce flood risk without unduly infringing on private property rights or usurping local decision making. Tackling this challenge would require significant adjustments in the flood insurance program, disaster aid policies and practices, and programs for structural and nonstructural measures and actions.

Steps Toward a Flood Policy Reorientation

Since Hurricane Katrina, Congress has conducted hearings (see Appendix A) and considered legislation on numerous aspects of federal flood programs and policies. Actions by many federal agencies shape the nation's flood risk management.[39] Legislative efforts since 2005 have largely proceeded by addressing individual programs or agencies, rather than through a comprehensive attempt to reorient flood policy. For example, in the 110th Congress, both the House and Senate passed a Flood Insurance Reform and Modernization Act (H.R. 3121 and S. 2284) aimed at changing the NFIP and FEMA's programs; this legislation was not enacted. The Water Resources Development Act of 2007 enacted numerous provisions related to Corps flood projects and programs. While implementation of WRDA 2007 provisions may shift the Corps' flood-related actions, few other changes to federal programs have been enacted.

In WRDA 2007, Section 2032 calls for the Administration to prepare a report by the end of 2009 describing flood risk and comparing regional risks. The report also is to assess the effectiveness of flood efforts and programs, analyze whether programs encourage development in flood-prone areas, and provide recommendations. The report's preparation, however, is delayed; the Corps has not received appropriations to prepare it.

Another provision in §2031 requires Corps feasibility studies to calculate a proposed flood damage reduction project's residual risk of flooding, loss of human life, and human safety. The benefit-cost calculations of the study also must include upstream and downstream impacts and give equitable consideration to structural and nonstructural alternatives.

Section 2031 of WRDA 2007 also called for the Secretary of the Army to update water resources planning guidance; the update would affect how Corps flood damage reduction projects are planned, evaluated, and selected. Sec. 2031 also stated:

NATIONAL WATER RESOURCES PLANNING POLICY.—It is the policy of the United States that all water resources projects should reflect national priorities, encourage economic development, and protect the environment by—

1. seeking to maximize sustainable economic development;
2. seeking to avoid the unwise use of floodplains and flood-prone areas and minimizing adverse impacts and vulnerabilities in any case in which a floodplain or flood-prone area must be used; and
3. protecting and restoring the functions of natural systems and mitigating any unavoidable damage to natural systems.

How this planning update and implementation of this policy statement may alter flood damage reduction and other water resources planning by federal agencies remains unknown.

Levee Reliability

Hurricane Katrina also brought national attention to the issue of levee and floodwall reliability and different levels of protection provided by flood damage reduction structures. Floodwall failures contributed to roughly half of the flood damages in New Orleans. A large percentage of locally built levees are poorly designed and maintained.

Section 9004 of WRDA 2007 required the Corps by 2009 to establish and maintain a national levee database. The database structure was completed; the process of populating the database with information on levees is ongoing. Section 9004 also requires the Corps to establish an inventory and inspect all federally owned and federally constructed levees. The provision also requires the Corps to establish an inventory of levees participating in the Corps' Repair and Inspection Program; the Corps may inspect these levees if requested by the owner. The Corps has completed an initial survey identifying 14,000 miles of Corps-owned, Corps-constructed, and RIP participating levees.

No federal program specifically regulates the design, placement, construction, or maintenance of nonfederal levees built by private individuals or by public entities such as levee districts. Section 9003 of WRDA 2007 created a National Committee on Levee Safety to make recommendations to Congress for a national levee safety program. WRDA 2007 also requires Corps planning to consider the risk that remains behind levees and floodwalls, evaluate upstream and downstream impacts, and equitably analyze structural and nonstructural alternatives. This provision put in statute requirements similar to direction in agency planning guidance.

How WRDA 2007 provisions and previous congressional direction (see Appendix A) are implemented and enforced, and whether the recommendations by the National Committee on Levee Safety (see Box 3) are pursued, may influence the nature of federal and local levee investments. However, levees represent only a portion of the nation's efforts at flood risk management.

REDUCING FLOOD RISK

Recommendations for how to improve flood policy abound. Figure 2 illustrates how different tools can combine to lower risk, but that some risk will always remain. Often following a significant flood or hurricane, changes are made to implement some tools and improve existing programs, but other tools and changes are not pursued. A comprehensive strategy to realign floodplain management would confront many challenges and require dramatic changes in how local, state, and federal government agencies and programs operate. One proposal for a national strategy was the 1986 Unified National Program for Floodplain Management by the Interagency Task Force on Floodplain Management. It laid out a four-part strategy for a balanced approach to floodplain management (see Box 4). Implementing the risk reduction tools in this strategy would realign government programs to reward behaviors that decrease flood risk. Use of these tools also would represent a policy choice to shift more of the long-term costs of staying or locating in flood-prone areas from the federal government to local communities and individuals.

**Box 3. Selected Recommendations in the
2009 Draft National Levee Safety Committee Report**

On January 15, 2009, the National Levee Safety Committee, established by WRDA 2007, released a draft of its report, *Recommendation for a National Levee Safety Program*.40 The report set out 20 principal recommendations, including:
Building and Sustaining Levee Safety Program in All States

- Design a levee safety program and delegate program responsibilities to states.
- Provide grants to assist in implementing the program.
- Establish a national levee rehabilitation, improvement, and flood

---

| mitigation fund to aid in improvement or removal of aging or deficient levees. |

Aligning Existing Federal Programs

- Align federal programs to provide incentives for good levee behavior.
- Mandate purchase of risk-based flood insurance in areas behind levees.

Comprehensive and Consistent National Leadership

- Establish a National Levee Safety Commission.
- Expand and maintain the National Levee Database.
- Develop and adopt national levee safety standards.
- Address growing concerns regarding liability of engineering firms and government agencies for damages resulting from levee failures.
- Develop a national levee safety training program.
- Develop a national public involvement and awareness campaign to communicate risk behind levees

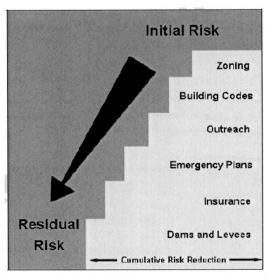

Source: Adapted from materials by D. Bollman, U.S. Army Corps of Engineers, *Raising the Standard: Post Flood Restoration Opportunities*, available at http://www.iwr.usace.army.mil/iltf/docs/Dorie_Final_DU_Presentation.pdf.

Figure 2. Multiple Tools Available To Reduce Flood Risk.

Box 4. Floodplain Management Strategy and Tools Proposed in the Unified National Program for Floodplain Management

Modify Susceptibility to Flood Damage and Disruption
1. Floodplain regulations
2. Development and redevelopment policies
3. Disaster preparedness
4. Disaster assistance
5. Floodproofing
6. Flood forecasting and waning systems and emergency

Modify Flooding

1. Dams and reservoirs
2. Levees, floodwalls, and dikes
3. Channel alterations
4. High flow diversions
5. Land treatment measures

Modify the Impact of Flooding on Individuals and the Community
1. Information and Education
2. Flood Insurance
3. Tax adjustments
4. Flood emergency measures
5. Post-flood recovery

Restore and Preserve the Natural and Cultural Resources of Floodplains

1. Floodplain, wetland, coastal barrier resources regulations
2. Development and redevelopment policies
3. Information and education
4. Tax adjustments

Source: Interagency Task Force on Floodplain Management, *A Unified National Program for Floodplain Management* (FEMA, March 1986).

Resilient Recovery

The 2008 Midwest flood and Hurricane Ike will not be the last riverine flood or coastal storm to affect these areas and devastate communities. See **Box 5** for a discussion of Hurricane Ike's impact on coastal communities and the challenge of recovery. To assist communities to rebuild in a more resilient manner, the Disaster Mitigation Act of 2000 (P.L. 106-390) requires that each state and community must have a mitigation plan to be eligible for certain disaster assistance. This planning requirement represents an initial step in improving the identification of risks; however, these plans have not translated into mitigation actions and assessment being incorporated into community comprehensive plans. There is no requirement for catastrophic recovery planning in communities that face significant risk, such as Galveston.[41] Such planning could assist recovery by vetting, prior to the disaster, preliminary needs, priorities, and plans for rebuilding. New Orleans after 2005 and the Texas coast, Cedar Rapids, and other severely affected Midwest communities after 2008 illustrate the challenge of undertaking an extensive recovery effort. It can be difficult to balance minimizing the disaster's disruption to the community and its economy and reestablishing a more resilient community.

**Box 5. The Challenge of Disaster Recovery:
The Case of Texas Coastal Communities after Hurricane Ike**

In September 2008, Hurricane Ike produced a damaging, destructive and deadly storm surge across the Gulf, affecting the eastern Texas and southwest Louisiana coasts most severely. At $27 billion and more than 100 deaths, Hurricane Ike is costlier and more deadly than the 2008 Midwest flood ($15.0 billion and 24 deaths). While more deadly, Hurricane Ike is closer to the 1993 Midwest flood ($30.2 billion and 48 deaths) in its economic impact. Hurricane Ike's disaster damages include not only coastal flood losses but also the effects of the storm's wind and other damage incurred as it tracked across Texas and the Midwest.

Many of the coastal communities damaged during Hurricane Ike face years of recovery. Much of the coastal residential damage was from storm surge, not wind. The storm generated storm surges between 10 and 13 feet in most of the area around Galveston, Texas; the surge in some areas reached 15 to 20 feet. Individuals, neighborhoods, and communities have many important surge-related rebuilding decisions.

> The threat of these communities being affected by another hurricane is significant; in less than 60 days during the summer of 2008, Hurricanes Dolly, Gustav, Ike and Tropical Storm Eduardo hit Texas.
>
> Thousands of families remain challenged to find affordable accommodations near their jobs and children's schools. After the storm, only 14 of 3,400 homes in the Texas town of Bridge City remained inhabitable. In Gilchrist, TX, only one home was left standing. Only 39% of flood damages were insured. Residents face difficult financial challenges and decisions regarding whether and how to rebuild.
>
> Business owners are faced with reinvestment decisions that depend on the collective decisions of property owners, their customer base, and municipalities. The City of Galveston's downtown historic strand was significantly damaged; up to 85% of the city's business base is gone. In the near-term, saltwater intrusion from the storm surge hurts agricultural production, and disruption to bay and coastal ecology hurts fishing and tourism. Nonetheless, the longterm regional economic development is likely to rebound. The petrochemical, fishing, and shipping industries remain tied to the Gulf of Mexico navigation infrastructure and fishery and oil resources.
>
> Each municipality is confronted with questions regarding development regulations, zoning ordinances, building codes, city planning, and utility and public safety service delivery. Local officials are faced with both recovery costs and a smaller base. Much of the storm damage was to public infrastructure further taxing local municipalities; repairs to most public infrastructure is eligible for some FEMA reimbursement.
>
> Source: CRS compiled primarily from FEMA's report, Hurricane Ike Impact report, December 2008, available at http://www.fema.gov/pdf/hazard/ hurricane/ 2008/ike/impact_report.pdf.

CONCLUDING REMARKS

The 2008 Midwest flood, Hurricane Ike, and Hurricane Katrina have been recent reminders of the nation's flood risk. These events have raised both concerns about the state of the nation's flood policies, programs, and infrastructure, and awareness of the tradeoff between the benefits and risks of developing flood-prone areas.

After the 1993 Midwest flood, Congress took several actions that departed from historic flood policies which relied heavily on structural solutions by providing more incentives and assistance for hazard mitigation. Nonetheless, many fundamental issues identified in reports following the 1993 flood remain today. Many federal, state, and local policies and programs continue to encourage floodplain development and use. Local-state-federal tensions over proper and respective roles and responsibilities continue to cloud resolution of difficult water resource issues and can slow recovery in disaster-affected communities. Flood damage reduction and mitigation projects are still largely authorized and implemented in piecemeal fashion, and water and related land use decisions and programs are rarely coordinated. Federal legislation enacted since 1993 generally has addressed individual programs or agencies, rather comprehensively dealing with the disparate federal policies, programs, and agencies influencing the nation's flood risk. In summary, although federal programs have improved through congressional and agency action since 1993, the fundamental direction and approach of national flood policies and programs remain largely unchanged.

APPENDIX A. CONGRESSIONAL FLOOD DIRECTION AND OVERSIGHT

Congress shapes how federal agencies implement their missions through authorization, appropriations, and oversight. To illustrate how Congress shapes federal agencies flood-related actions through legislative direction, Table A-1 provides a list of the direction that Congress has given to the Corps on how the agency should implement its flood damage reduction mission and conduct its flood studies and projects. How to use this direction in guiding implementation can be challenging when one provision of law may conflict with another. Table A-2 is a listing of flood-focused congressional hearings between the 1993 Midwest flood and 2008; the table illustrates the role and focus of congressional oversight. Table A-3 is a list of flood-focused reports by the GAO from 1993 through 2008; GAO reports investigate how the federal government spends taxpayer dollars in order to assist Congress in meeting its constitutional responsibilities and to help improve the performance and ensure the accountability of the federal government.

Table A-1. Selected Congressional Direction on Corps Flood Damage Reduction Efforts

Topic	Section of Law	Summary of Effect	U.S. Code
Federal Involvement in Flood Damage Reduction Projects			
Flood	§1 of Flood Control Act of 1936 (49 Stat. 1570)	Declared flood control a proper federal activity and that the federal government should participate in the flood control improvements of navigable waters or their tributaries if the benefits are in excess of the costs and if the lives and social security of people are otherwise adversely affected.	33 USC 701a
Shore Protection	§2018 of WRDA 2007	Established that it is the policy of the United States to promote beach nourishment for flood and storm damage reduction and related research, including beach restoration and periodic nourishment for 20 years on a comprehensive and coordinated basis by federal, state, local, and private entities.	33 USC 426e-1
Nonstructural Measures in Corps Flood Damage Reduction Projects			
Flood Risk, Adjacent Impacts, and Nonstructural Alternatives	§2033 of WRDA 2007	Required that a feasibility study ensure equitable analysis of structural and nonstructural alternatives.	33 USC 2282a
Nonstructural Flood Alternatives	§219 of WRDA 1999	Directed that benefits of nonstructural alternatives be calculated using methods similar to those for structural projects, and that double counting of benefits should be avoided.	33 USC 2318
Nonstructural Alternatives	§905 of WRDA 1986	Required feasibility reports to describe a nonstructural alternative to the recommended plan when it does not include significant nonstructural features.	33 USC 2282
Nonstructural Flood Alternatives	§73 of WRDA 1974 (P.L. 93-251)	Required consideration of nonstructural alternatives for flood damage prevention or reduction during planning.	33 USC 701b-11
Evacuation in Lieu of Levees	§3 of Flood Control Act of 1938 (52 Stat. 1216)	Allowed the Chief of Engineers to modify the plan of any authorized flood control project if construction costs can be substantially reduced by the evacuation of a portion or all of the area proposed to be protected and to use the amount saved toward the evacuation costs, including rehabilitation of the persons evacuated.	33 USC 701i

Table A-1. (Continued)

Topic	Section of Law	Summary of Effect	U.S. Code
Analysis of Corps Flood Projects			
Flood Risk, Adjacent Impacts	§2033 of WRDA 2007	Required that a feasibility report include as part of the calculation of benefits and costs the residual risk of flooding, residual risk of loss of life and human safety, and upstream or downstream impacts.	33 USC 2282a
Flood Control Integration	§216 of WRDA 1999 (P.L. 106-53)	Required the Secretary of the Army to coordinate with FEMA and other federal agencies to ensure that flood control projects and plans are complementary and integrated.	33 USC 709a
Exclusion of Floodplain Development	§308 of WRDA 1990	Directed that the Secretary, in justifications for new federal projects, cannot consider benefits from protecting new or substantially improved structures built in the 100-year floodplain after July 1991.	33 USC 2318
Uneconomic Flood Increments	§903 of WRDA 1986	Authorized the Secretary to include flood control features that do not have national economic development benefits greater than costs if the nonfederal interest pays for the element until the remaining costs equal the national economic benefit.	Not codified. 100 Stat. 4184
Flood Measures	§914 of WRDA 1986	Allowed the Secretary to consider flood damage reduction measures without regard for flood frequency, drainage area, or runoff amount, so long as the federal share is less than $3 million.	33 USC 2289
Watershed Analysis for Flood Control	§3 of Flood Control Act of 1917 (39 Stat. 950)	Required surveys for flood control to include a comprehensive study of the watershed.	33 USC 701
General Water Policies Affecting Flood Damage Reduction Projects			
Water Projects	§2031 of WRDA 2007 (P.L. 110-114)	Established as U.S. policy that all water resources projects reflect national priorities, encourage economic development, and protect the environment by maximizing sustainable economic development, avoiding unwise use of flood-prone areas and minimizing adverse impacts of floods, protecting and restoring natural system functions, and mitigating unavoidable natural system damage.	42 USC 1962-3

Topic	Section of Law	Summary of Effect	U.S. Code
Water Policy	§2 of Water Resources Planning Act of 1965 (P.L. 89-80)	Declared that congressional policy is to encourage conservation, development, and utilization of water and related land resources on a comprehensive and coordinated basis by the federal government, states, localities, and private enterprise with the cooperation of all affected and others.	42 USC 1962
Water Policy (including state deference)	§1 of Flood Control Act of 1944 (58 Stat. 887)	Declared that congressional policy is to recognize rights and interests of the states in water resources development, to preserve and protect potential uses, to facilitate project consideration on a comprehensive and coordinated basis, and to limit navigation works to those with substantial benefit which can be operated consistently with appropriate and economic water uses.	33 USC 701-1

Source: CRS, compiled from public laws and assisted by Chapter 2 of Corps, *Digest of Water Policies and Authorities* (EP 1165-2-1, July 1999).

Table A-2. Flood-Focused Congressional Hearings from Summer 1993 through 2008

Hearing Date by Theme	Hearing Title	Committee and Subcommittee
Flood Policy		
Oct. 27, 2005	Reducing Hurricane and Flood Risk in the Nation	House Transportation and Infrastructure (T&I) Subcommittee on Water Resources and Environment
May 26, 1994; July 20, 1994	Floodplain Management and Flood Control	Senate Environment and Public Works (EPW)
Levees		
May 8, 2007	National Levee Safety and Dam Safety Programs	Joint hearing: House T&I Subcommittee on Economic Development, Public Buildings, and Emergency Management; and Subcommittee on Water Resources and Environment
April 6, 2006	H.R. 4650, the National Levee Safety Program Act	House T&I Subcommittee on Water Resources and Environment
April 10, 1997	Flood Control Projects and ESA	House Resources
Hazard Mitigation and Floodplain Management		

Table A-2. (Continued).

Hearing Date by Theme	Hearing Title	Committee and Subcommittee
June 24, 2008	Comprehensive Watershed Management Planning	House T&I Subcommittee on Water Resources and Environment
Flood-Related Disaster Aid and Response		
Feb. 16, 2000	Flood Water Rescue	House T&I Subcommittee on Oversight, Investigations, and Emergency Management
March 26, 1998	Federal Cost of Disaster Assistance	House T&I Subcommittee on Water Resources and Environment
July 14, 1993	Agricultural Disaster Relief	House Agriculture
Flood Insurance and Mapping		
April 2, 2008	National Flood Plain Remapping	House T&I Subcommittee on Economic Development, Public Buildings, and Emergency Management
July 17, 2007	H.R. 920, the Multiple Peril Insurance Act of 2007	House Financial Services Subcommittee on Housing and Community Opportunity
June 12, 2007	Flood Insurance Reform and Modernization Act of 2007	House Financial Services Subcommittee on Housing and Community Opportunity
June 12, 2007	National Flood Insurance Program: Issues Exposed by the 2005 Hurricanes	Joint Hearing: House Financial Services Subcommittee on Oversight and Investigations and House Homeland Security Subcommittee on Management, Investigations, and Oversight
Feb. 28, 2007	Insurance Claims Payment Process in the Gulf Coast After the 2005 Hurricanes	House Financial Services Subcommittee on Oversight and Investigations
Aug. 15, 2006	Look at the National Flood Insurance Program and Flood Mitigation Efforts: Is Bucks Country, Pennsylvania Ready for Another Flood?	House Financial Services
May 8, 2006	FEMA's Floodplain Map Modernization: A State and Local Perspective	House Government Reform Subcommittee on Regulatory Affairs
Oct. 20, 2005	Management and Oversight of the National Flood Insurance Program	House Financial Services Subcommittee on Housing and Community Opportunity
Oct. 2, 2005; Jan. 25, 2006; and Feb. 2, 2006	Future of the National Flood Insurance Program	Senate Banking, Housing, and Urban Affairs

Hearing Date by Theme	Hearing Title	Committee and Subcommittee
Aug. 17, 2005	Look at the National Flood Insurance Program: Is Ohio Ready for a Flood?	House Financial Services Subcommittee on Housing and Community Opportunity
June 12, 2005	Flood Map Modernization and the Future of the National Flood Insurance Program	House Financial Services Subcommittee on Housing and Community Opportunity
April 14, 2005	Review and Oversight of the National Flood Insurance Program	House Financial Services Subcommittee on Housing and Community Opportunity
March 25, 2004	National Flood Insurance Repetitive Losses	Senate Banking, Housing, and Urban Affairs Subcommittee on Economic Policy
April 1, 2003	National Flood Insurance Program: Review and Reauthorization	House Financial Services Subcommittee on Housing and Community Opportunity
July 19, 2001	National Flood Insurance Program and Repetitive Loss Properties	House Financial Services Subcommittee on Housing and Community Opportunity
Oct. 27, 1999	National Flood Insurance Program	House Banking and Financial Services Subcommittee on Housing and Community Opportunity
June 24, 1993	Status of the National Flood Insurance Program	House Banking, Finance, and Urban Affairs Subcommittee on Consumer Credit and Insurance
May 27, 1993	Insurance Availability in Communities at Risk of Natural Disaster	House Banking, Finance, and Urban Affairs Subcommittee on Consumer Credit and Insurance
Hurricane Katrina & Coastal Louisiana		
April 18, 2006	Field Hearing: Oversee the Ongoing Rebuilding and Restoration Efforts of Hurricane and Flood Protection by the Army Corps of Engineers	Senate EPW
Dec. 15, 2005	Hurricane Katrina: Who's in Charge of the New Orleans Levees?	Senate Homeland Security and Government Affairs Committee
Nov. 17, 2005	Evaluate the Degree to Which the Preliminary Findings on the Failure of the Levees Are Being Incorporated into the Restoration of Hurricane Protection	Senate EPW
Nov. 9, 2005	Comprehensive and Integrated Approach to meet the Water Resources Needs in the Wake of Hurricanes Katrina and Rita	Senate EPW
Nov. 9, 2005	Coastal Louisiana Hurricane Protection Project	Senate EPW
Nov. 2, 2005	Hurricane Katrina: Why Did the Levees Fail?	Senate Homeland Security and Government Affairs Committee

Table A-2. (Continued).

Hearing Date by Theme	Hearing Title	Committee and Subcommittee
Nov. 2, 2005	Second in a Series of Two Hearings to Discuss the Response to Hurricane Katrina	Senate EPW
Oct. 20, 2005	Expert Views on Hurricane and Flood Protection and Water Resources Planning for a Rebuilt Gulf Coast	House T&I Subcommittee on Water Resources and Environment
Oct. 6, 2005	Roles of the Environmental Protection Agency, the Federal Highway Administration and the Army Corps of Engineers as they Relate to Katrina and the Ongoing Recovery (First in a Series of Two)	Senate EPW
Sept. 29, 2005	Hurricane Katrina: Assessing the Present Environmental Status	House Energy and Commerce Subcommittee On Environment and hazardous Materials
Midwest Floods		
July 23, 2008	The Midwest Floods: What Happened and What Might Be Improved for Managing Risk and Responses in the Future	Senate EPW
Feb. 22-24, 1994	Condition of Agricultural Land Damaged by the Midwest Flood	House Agriculture Subcommittee On General Farm Commodities and Subcommittee On Environment, Credit, and Rural Development
Nov. 19, 1993	SBA Disaster Assistance Programs	House Small Business
Nov. 9, 1993	Federal Response to the Midwest Floods of 1993	Senate EPW
Oct. 28, 1993	Federal Response to Midwest Flooding	House Public Works and Transportation Subcommittee. On Investigations and Oversight
Oct. 27, 1993	Midwest Floods of 1993: Flood Control and Floodplain Policy and Proposals	House Public Works and Transportation Subcommittee on Water Resources and Environment
Sept. 23, 1993	Effect of Midwest Flooding on Rail Transportation	House Energy and Commerce Subcommittee on Transportation and Hazardous Materials
Sept. 14-15, 1993	National Flood Insurance Reform Act of 1993: S. 1405	Senate Banking, Housing, and Urban Affairs Subcommittee on Housing and Urban Affairs
July 16, 1993	Flood and Disaster Relief in the Midwest	Senate Agriculture, Nutrition, and Forestry

Federal Flood Policy Challenges 39

Hearing Date by Theme	Hearing Title	Committee and Subcommittee
Other Geographically Specific Flood Issues		
Aug. 23, 2008	Hearing on the Small Business Administration's Response to the July 5 Floods in Beaver County, Pennsylvania	House Small Business Subcommittee on Investigations and Oversight
Nov. 1, 2007	Impact of the Flood Control Act of 1944 on Indian Tribes Along the Missouri River	Senate Indian Affairs
Nov.28, 2006	FEMA's Response to the Rockford Flood	House Small Business
Sept. 22, 1998	Coastal Barrier Resources System: Pumpkin Key, Florida	Senate EPW
May 27, 1998	Field Hearing on Proposed Modifications of Folsom Dam	House Resources/Subcommittee on Water and Power
Oct. 23, 1997	Flood Control at Devils Lake, North Dakota	Senate EPW
March 19, 1997	Recent Flooding in California	House T&I Subcommittee on Water Resources and Environment

Source: Information compiled by Lynn J. Cunningham, Information Research Specialist, CRS Knowledge Services Group.

Notes: CRS identified these hearings using flood-related keywords in the hearing title; flood issues may have been discussed during other congressional hearings during this period.

Table A-3. Flood-Focused GAO Reports from Summer 1993 through 2008

Report Date by Theme	GAO Report Title
Flood Policy	
Jan. 1, 2003	Major Management Challenges and Program Risks: Federal Emergency Management Agency, GAO-03-113
Levees	
	None identified.
Hazard Mitigation and Floodplain Management	
Aug. 22, 2007	Natural Hazard Mitigation: Various Mitigation Efforts Exist, but Federal Efforts Do Not Provide a Comprehensive Strategic Framework, GAO-07-403
March 19, 2007	Coastal Barrier Resources System: Status of Development That Has Occurred and Financial Assistance Provided by Federal Agencies, GAO-07-356

Table A-3. (Continued).

Report Date by Theme	GAO Report Title
April 21, 2003	Agricultural Conservation: USDA Needs to Better Ensure Protection of Highly Erodible Cropland and Wetlands, GAO-03-418
Dec. 20, 2002	Results-Oriented Management: Agency Crosscutting Actions and Plans in Border Control, Flood Mitigation and Insurance, Wetlands, and Wildland Fire Management, GAO-03-321 (also listed under Flood Insurance and Mapping)
Aug. 4, 1999	Disaster Assistance: Opportunities to Improve Cost-Effectiveness Determinations for Mitigation Grants, RCED-99-236
Flood-Related Disaster Aid and Response	
Aug. 4, 1999	Disaster Assistance: Opportunities to Improve Cost-Effectiveness Determinations for Mitigation Grants, RCED-99-236 (also listed under Hazard Mitigation and Floodplain Management)
Flood Insurance and Mapping	
June 16, 2008	National Flood Insurance Program: Financial Challenges Underscore Need for Improved Oversight of Mitigation Programs and Key Contracts, GAO-08-437
April 25, 2008	Natural Catastrophe Insurance: Analysis of a Proposed Combined Federal Flood and Wind Insurance Program, GAO-08-504
Dec. 28, 2007	National Flood Insurance Program: Greater Transparency and Oversight of Wind and Flood Damage Determinations Are Needed, GAO-08-28
Nov. 26, 2007	Natural Disasters: Public Policy Options for Changing the Federal Role in Natural Catastrophe Insurance, GAO-08-7
Sept. 5, 2007	National Flood Insurance Program: FEMA's Management and Oversight of Payments for Insurance Company Services Should Be Improved, GAO-07-1078
March 16, 2007	Climate Change: Financial Risks to Federal and Private Insurers in Coming Decades Are Potentially Significant, GAO-07-285
Dec. 15, 2006	National Flood Insurance Program: New Processes Aided Hurricane Katrina Claims Handling, but FEMA's Oversight Should Be Improved, GAO-07-169
Oct. 18, 2005	Federal Emergency Management Agency: Improvements Needed to Enhance Oversight and Management of the National Flood Insurance Program, GAO-06-119
March 31, 2004	Flood Map Modernization: Program Strategy Shows Promise, but Challenges Remain, GAO-04-417

Federal Flood Policy Challenges 41

Report Date by Theme	GAO Report Title
Dec. 20, 2002	Results-Oriented Management: Agency Crosscutting Actions and Plans in Border Control, Flood Mitigation and Insurance, Wetlands, and Wildland Fire Management, GAO-03-321 (also listed under Hazard Mitigation and Floodplain Management)
June 21, 2002	Flood Insurance: Extent of Noncompliance with Purchase Requirements Is Unknown, GAO-02-396
Hurricane Katrina & Coastal Louisiana	
Dec. 31, 2007	Army Corps of Engineers: Known Performance Issues with New Orleans Drainage Canal Pumps Have Been Addressed, but Guidance on Future Contracts Is Needed, GAO-08-288
Dec. 14, 2007	Coastal Wetlands: Lessons Learned from Past Efforts in Louisiana Could Help Guide Future Restoration and Protection, GAO-08-130
June 29, 2007	Preliminary Information on Rebuilding Efforts in the Gulf Coast, GAO-07-809R,
June 25, 2007	Hurricane Katrina: EPA's Current and Future Environmental Protection Efforts Could Be Enhanced by Addressing Issues and Challenges Faced on the Gulf Coast, GAO-07-651
May 23, 2007	U.S. Army Corps of Engineers' Procurement of Pumping Systems for the New Orleans Drainage Canals, GAO-07-908R
Sept. 6, 2006	Hurricane Katrina: Strategic Planning Needed to Guide Future Enhancements Beyond Interim Levee Repairs, GAO-06-934
Midwest Floods	
Aug. 7, 1995	Midwest Flood: Information on the Performance, Effects, and Control of Levees, RCED-95-125
Other Geographically Specific Flood Issues	
April 16, 2007	IRS Emergency Planning: Headquarters Plans Supported Response to 2006 Flooding, but Additional Guidance Could Improve All Hazard Preparedness, GAO-07-579
Dec. 12, 2003	Alaska Native Villages: Most Are Affected by Flooding and Erosion, but Few Qualify for Federal Assistance, GAO-04-142
Oct. 27, 2003	Corps Of Engineers: Improved Analysis of Costs and Benefits Needed for Sacramento Flood Protection Project, GAO-04-30
Dec. 15, 1999	Food and Drug Administration Facility: Requirements for Building on a Floodplain Met, GGD-00-17
Oct. 2, 1996	Bureau of Reclamation: An Assessment of the Environmental Impact Statement on the Operations of the Glen Canyon Dam, RCED-97-12

Table A-3. (Continued).

Report Date by Theme	GAO Report Title
Other	
June 8, 2007	Weather Forecasting: National Weather Service's Operations Prototype Needs More Rigorous Planning, GAO-07-650
May 15, 2002	U.S. Army Corps of Engineers: Scientific Panel's Assessment of Fish and Wildlife Mitigation Guidance, GAO-02-574
July 9, 2001	Federal Emergency Management Agency: Status of Achieving Key Outcomes and Addressing Major Management Challenges, GAO-01-832
April 2, 1996	Lands Managed by the Corps of Engineers, RCED-96-101R
Aug. 12, 1993	Water Resources: Factors That Lead to Successful Cost Sharing in Corps Projects, RCED-93-114

Source: Information compiled by the Wayne A. Morrissey, Information Research Specialist, CRS Knowledge Services Group.

Notes: CRS identified these reports using flood-related keywords; flood issues may have been discussed in other GAO reports not listed in the table.

APPENDIX B. ANALYSIS OF THE UPPER MISSISSIPPI RIVER COMPREHENSIVE PLAN

WRDA 1999 authorized the Upper Mississippi River Comprehensive Plan (UMRCP). The UMRCP final report, which is dated June 2008, was transmitted to Congress on January 15, 2009. The UMRCP was conducted as a preliminary study, similar to the level of detail in a Corps reconnaissance study. The UMRCP final report and supporting documents are not at the level of detail of a feasibility study, which typically informs decision-making on construction authorization.

The UMRCP Preferred (But Not Recommended) Plan

The UMRCP final report states "additional authority to implement the Comprehensive Plan is not being recommended nor requested at this time based upon the [national economic development] evaluation of alternative plans." Nonetheless, the UMRCP final report did identify a preferred alternative from among the fourteen analyzed; the alternatives were evaluated on multiple criteria, including environmental, social, and regional benefits. The report presented one "no action" alternative and thirteen other alternatives that would provide 500-year urban protection. These thirteen varied primarily on the level of protection and type of flood damage reduction actions taken in agricultural areas; the alternatives ranged from increase in existing protection to 500-year protection for agricultural areas.

The preferred alternative, known as Plan H, would provide a 500-year level of flood protection along the length of the mainstem of the Mississippi and Illinois Rivers (but not other tributaries) and ecosystem restoration benefits. The preferred plan would protect urban areas and towns with 500-year levees; for agricultural areas it would provide 500-year levees except where buyouts would be more cost effective.[42] The UMRCP final report indicates that up to 39 levee districts would be bought out while 144 would have levees raised to 500-year protection. If buyouts of districts are implemented, there would be opportunities to pursue ecosystem restoration actions. The total initial cost for Plan H would be $4.42 billion — $3.97 billion for flood damage reduction construction, and $460 million for ecosystem restoration; these costs do not include operation, maintenance, and rehabilitation.

Mixed Recommendations on How to Proceed

The UMRCP final report indicates that none of the alternatives studied would meet the current economic test for federal participation of the plan's national benefits exceeding costs. Current guidelines exclude regional benefits from these calculations because regional benefits are viewed as transfers from one region to another, and do not produce national gains.

The thirteen UMRCP alternatives analyzed (excluding the no action alternative) had costs from $3 billion to $9 billion and benefit-cost ratios ranging from 0.03 to 0.07 for the national economic development benefits. For Corps projects, other than ecosystem restoration projects, a national benefit-cost ratio greater than 1.0 generally is used in gauging the economic attractiveness of the federal investment, consistent with the direction in the Flood Control Act of 1936.

Congress is faced with deciding how to proceed given differing recommendations. Consistent with the UMRBC final report, the Assistant Secretary of the Army (Civil Works) in his January 2009 transmittal letter to Congress stated that "recommendations for implementation of a specific plan based on a reconnaissance level of detail is premature."[43] The Assistant Secretary instead recommended intermediate steps — expanding the UMRCP to include

Mississippi River tributaries, conducting cost-shared studies of the reconstruction needs for the basin's existing flood damage infrastructure (where a federal interest is identified), and conducting a study of flood protection for critical transportation infrastructure such as bridge approaches and railroads.

Earlier in August 2008, the Mississippi River Commission voted to support implementing the preferred alternative.[44] The Commission believed that the full benefits of implementing the preferred alternative were not adequately measured with the current Corps project planning guidelines.[45]

In 2008, the then-Governor of Illinois and the then-Governor of Missouri wrote letters of support for Plan M. Plan M at a total cost of $6.88 billion would provide 500-year protection without the option for agricultural district buyouts and without trying to minimize the impacts in the lower basin (i.e., Plan M would increase the height of floodwaters below St. Louis). Plan M would provide less ecosystem restoration opportunities than Plan H.

A Plan with Limited Scope and Detail

Although the study authorization was labeled as comprehensive and inclusive of some navigation maintenance and habitat management considerations, the authorized flood study did not fully integrate navigation, flood, and ecosystem management as recommended in the 1994 Galloway report. Instead, the Corps studied and obtained construction authorization for navigation and ecosystem restoration actions (in Title VIII of WRDA 2007, P.L. 110-114) separately from the flood plan.

Due to the large study area for the flood plan, the Corps chose not to perform a comprehensive watershed analysis encompassing the entire 185,000 square miles, instead it limited the study to the Mississippi and Illinois River floodplain encompassing 4,000 square miles, and the only tributary that was included was the Illinois River. The Missouri River and smaller tributaries were excluded.

For the comprehensive flood plan, the Corps identified preliminary alternatives and scoped out the federal interest in the effort; the level of detail of the plan is compared to a Corps reconnaissance study. Therefore, the UMRCP final report and supporting documentation are not at the level of detail typically used to inform congressional decision-making regarding construction authorization.

The analyses used to support the UMRCP (e.g., counting as benefits the increased development opportunities behind levees[46]), the scoping of the study and the selection of alternatives studied, and whether nonstructural alternatives and enhanced floodplains were given equal consideration are some of the items that may be scrutinized as the final report is discussed. For example, the hydrology and hydraulics analysis supporting the UMRCP final report did not account for the effects on precipitation, runoff, and river crests from future changes in land use, population, and climate.[47] Moreover, the scope of the UMRCP final report leaves out much of the areas most severely affected in 2008.

Visions of the Future Floodplain

Whether Plan H, particularly the raising of most agricultural levees to a significantly higher level of protection than currently available, contrasts with the vision of the future floodplain described in the Galloway report likely will be debated. The Galloway report stated:

> Urban centers whose existence depends on a river for commerce or whose locational advantage is tied historically to a floodplain would be protected from the ravages of devastating floods by means of levees, floodwalls, upstream reservoirs, or floodwater storage in managed upland and floodplain natural areas. Sections of communities with frequently flooded businesses or homes would become river-focused parks and recreation areas as former occupants relocated to safer areas on higher ground. In areas outside of these highly protected communities, where land elevation provided natural protection from floods, state and local officials would control new construction by requiring it to be at elevation well out of harm's way. Those who were at risk in low-lying areas would be relocated, over time, to other areas. ... Outside of the urban areas, industry would protect its own facilities against major floods. Critical infrastructure, such as water and wastewater treatment plants, power plants, and major highways and bridges would be either, elevated out of the flood's reach or protected against its ravages. Much of the infrastructure, as well as the homes, businesses, and agricultural activities located behind lower levees, would be insured against flooding through participation in commercial or federally supported insurance programs.[48]

The potential role of higher mainstem levees in increasing risk because of their encouragement of floodplain development and reduction in flood storage is an active part of the debate over the future of the basin's floodplains. The experience of extreme floodwaters along Mississippi River tributaries in the 2008 flood and differing visions for the future of the upper Mississippi River basin floodplains may be central to the debate about how to proceed with reducing flood risk in the Midwest and the UMRCP.

Regional Development and the Federal Role

According to the economic analysis used for the development of the UMRCP final report, the regional economic benefit of an alternative similar to Plan H would be $27.1 billion. The majority of regional benefits (79%) cluster in Illinois, with Iowa and Missouri receiving most of the remaining benefits. Therefore, regional stakeholders, particularly in Illinois, Iowa, and Missouri, may view plans, like Plan H and Plan M, as attractive investments. The majority of those regional economic benefits ($20.5 billion) are due to the increase in economic development behind the higher levees. Plan H potentially would open to development up to 215,775 acres. This potential for expanded economic development behind levees raises concerns regarding the residual risk behind levees and the evaluation of that risk in selecting Plan H as the preferred alternative. That is, it remains unclear the extent to which the flood risk reduction benefits of Plan H may be offset by the residual risk of more development behind levees. The methodology used in developing the study appears to be more similar to the traditional Corps flood damage reduction study, than a flood risk reduction study.

As well as noting that Plan H has not been thoroughly vetted with the public and stakeholders, the UMRCP final report stated:

> There is likely to be limited Federal interest, based upon current guidance, in plan implementation by Federal agencies.... Regional or national oversight (e.g., the Mississippi River Commission) would be required to ensure the plan functions as a system over the implementation and operation phases of the project and project priorities are established to reflect the changing systemic needs.... The States of Illinois, Iowa, and Missouri need to agree on the plan and plan implementation to insure the plan is acceptable. The Corps could provide facilitation and technical support to this effort.[49]

In effect, the UMRBC final report is identifying that the states could choose to further develop then implement one of the alternatives studied without significant federal leadership or funding.

End Notes

[1] These estimates are a lower bound from the January 1997 FEMA report, *FEMA's Multi-Hazard Identification and Risk Assessment (MHIRA)*, available at http://www.fema.gov/library/viewRecord.do?id=2214. The magnitude of flood events traditionally has been measured

by recurrence intervals, or the likelihood that a flood of a particular size occurring during any 10-, 50-, 100-, or 500-year period. Respectively, these events have a 10%, 2%, 1%, and 0.2% chance of being equaled or exceeded during any year.

[2] Information in this paragraph is from U.S. Army, *Fiscal Year 2008 United States Army Annual U.S. Army Corps of Engineers—Civil Works Financial Statement*.

[3] In this chapter, the term levees is used broadly to encompass both levees and floodwalls. Levees are broad, earthen structures, while floodwalls are concrete and steel walls, built atop a levee or in lieu of a levee. Floodwalls are often used in urban areas because they require less land than levees.

[4] The Corps performed an after-action report on its emergency response to the 2008 floods, but it did not conduct an engineering analysis to confirm whether the levees performed as designed.

[5] The then-General Accounting Office (renamed the Government Accountability Office, GAO) found that, according to three modeling simulations, the levees in the basin increased the height of water in the 1993 flood. For more information see GAO, *Midwest Flood: Information on the Performance, Effects, and Control of Levees* (GAO/RCED 95-125, Aug. 1995). Hereafter referred to as 1995 GAO Midwest Flood report.

[6] Precipitation from January to June of 2008 exceeded levels for the same period in 1993. After June the severe precipitation largely subsided for the remaining summer months. In contrast, the major rains in 1993 occurred in June and July, resulting in the most significant flooding occurring in the later summer months.

[7] On the Iowa River, water flowed over the spillway at the Corps' Coralville Reservoir for only the third time since the reservoir began operation in 1958; the other two times were during the Midwest flood of 1993. Unprecedented flooding occurred in Columbus Junction, Iowa City and Coralville. The flood set the record at Columbus Junction at 32.49 feet. Flood stage is 19 feet. The flood's record crest at Iowa City was 31.53 feet. Flood stage is 22 feet. For more information, see the National Weather Service website at http://www.crh.noaa.gov/images/dvn/downloads/fall08.pdf and the U.S. Geological Survey website at http://ia.water.usgs.gov/flood08/.

[8] The effects of the 1993 storms were exacerbated by preexisting saturated soils in the basin. The fall of 1992 was wet, saturating soils and raising stream levels. Winter rain and snow contributed to the nearly saturated soil conditions forcing spring precipitation and snowmelt, normally able to soak into the ground, to run off into streams and rivers. Heavy rainfall in late March fed directly into the headwaters of the Mississippi River. With the saturated soils, the precipitation in June, July, and August flowed directly to streams.

[9] Damage amounts are in normalized 2007 dollars. National Climate Data Center, National Oceanic and Atmospheric Administration, *1980-2009 Billion Dollar U.S. Weather Disasters*, available at http://www.ncdc.noaa.gov/img/reports/billion/disasters2009.pdf. Hereafter referred to as NOAA 1980-2009 Billion Dollar U.S. Weather Disasters.

[10] Corps, *Upper Mississippi River Comprehensive Plan: Final Report June 2008*, available at http://www.mvr.usace.army.mil/PublicAffairsOffice/MidwestFlooding2008/UMRCPFinalReport-17Jun08.pdf. Hereafter referred to as UMRCP final report.

[11] The flood carried away more than 600 billion tons of top soil and deposited great amounts of sand and silt on valuable farm land. In large areas inundated by the flood, the harvest of 1993 was a total loss. Although most farmers recovered and had good harvests in 1994, some farmers were affected through the 1994 harvest.

[12] Information in this paragraph is from GAO Midwest Flood Aug. 1995 report and UMRCP final report.

[13] Damage amounts are in normalized 2007 dollars. NOAA 1980-2009 Billion Dollar U.S. Weather Disasters.

[14] Flooding forced the closure of I-80, I-380, and US 34. On I-80, flood waters from the Cedar River flowed over the interstate resulting in its closure between mile markers 265 and 267 (between Davenport and Iowa City) from June 6 through the 12th. The official detour route added 115 miles to the normal east-west route across the state. Flooding from Coralville Reservoir resulted in the closure of I-380 between Iowa City and Cedar Rapids; the detour route added 272 miles to the normal route.

[15] Upper Mississippi River Basin Association, *Position of the Upper Mississippi River Basin Association on Flood Response and Recovery in the Wake of the 2008 Flooding: An Update to UMRBA's 1993 Flood Statement*, (St. Paul: Sep. 2008), available at http://www.umrba.org/publications/fp/flood9-17-08.htm. [16] 1995 GAO Midwest Flood report.

[17] J. D. Hipple et al., "Development in the Upper Mississippi Basin: 10 years after the Great Flood of 1993 " *Landscape and Urban Planning* (72, 2005, pp.313-323).

[18] Ibid.

[19] Information in this paragraph is from "U.S. Geological Survey—Rainfall-River Forecast Summit" in *Interagency Task Force, Raising the Standard*, Oct./Nov. 2008 newsletter, available at http://www.iwr.usace.army.mil/ILTF/docs/ILTF_Newsletter_OctNov_08.pdf.

[20] Testimony by Brigadier General Michael J. Walsh, Army Corps of Engineers Mississippi Valley Division Commander, Senate Environment and Public Works Committee hearing on the Midwest Flood of 2008, July 23, 2008.

[21] RIP is a Corps program that serves three main functions. It provides for inspections of flood-related works (including levees and flood control dams); and it provides assistance to repair these works if damaged by a flood or other damaging events (e.g., earthquake). The program also rehabilitates federally authorized and constructed hurricane or shore protection projects (including beach nourishment) damaged by an extraordinary storm (i.e., a storm that, due to length or severity, causes significant damage to a project). Rehabilitation generally is cost shared at 80% federal and 20% nonfederal. The Corps' Chief of Engineers, when requested by the nonfederal sponsor, is authorized to implement nonstructural alternatives to repair; the Corps may bear up to 100% of these costs, subject to limitations. RIP assistance is limited to restoration to pre-disaster conditions and level of protection. Only flood works and hurricane/storm projects that are active in the program at the time of the damaging event are eligible for assistance. RIP participation requirements include that the levee have a public sponsor and be deemed through regular inspections to be properly constructed and maintained. Another participation criterion for levees and floodwalls constructed is provision of at least a 10-year protection for urban areas or a year level of protection for agricultural areas.

[22] The memo is available at http://www.iwr.usace.army.mil/ILTF/docs/ OMB%20 CEQ%20Directive.pdf.

[23] For more information, see CRS Report R40073, *FEMA Funding for Flood Map Modernization*, by Wayne A. Morrissey.

[24] Some concerns also have been raised about the aging of these works. Many levees were privately built between 1880 and 1920, then later upgraded. On average, the last major upgrades occurred nearly 50 years ago.

[25] In the early 1980s, President Reagan dissolved most large-scale river basin commissions. The commissions had received mixed reviews. They raised state concerns about federal planning that could influence water supply allocation, which historically has been deferred to the states. Some water resource stakeholders have argued that the dissolution of the commissions has

resulted in a planning gap for basin-scale integrated water and related-land resource management.

[26] The Commission's statutory authority is the 1879 Mississippi River Commission Act (Chap. 42, 21 Stat. 37 (1879)).

[27] Hereafter referred to as 1994 Galloway report, available at http://www.floods

[28] UMRCP final report available at http://www.mvr.usace.army.mil/Public AffairsOffice/ MidwestFlooding2008/ UMRCPFinalReport-17Jun08.pdf.

[29] Ibid.

[30] The Mississippi River Commission press release is available at http://www.mvd.usace.army.mil/offices/pa/releases/ 2008/RelMRC0801.pdf. Half of the Commission's members are Corps officers.

[31] Letter to Hon. James M. Inhofe, Ranking Member Committee on Environment and Public Works, from Assistant Secretary of the Army (Civil Works) John Paul Woodley, Jr., transmitting the UMRPC final report, Jan. 15, 2009.

[32] National Water Commission, *Water Policies for the Future: Final Report to the President and to the Congress of the United States* (Washington: GPO, 1973); Interagency Task Force on Floodplain Management, *A Unified National Program for Floodplain Management* (FEMA, March 1996), available at http://www.fema.gov/hazard/flood/pubs/lib100.shtm.

[33] Information in this paragraph is drawn from G. Bucco, *Lessons Learned* available at http://www.dnr.state.ne.us/ floodplain/PDF_Files/Lessons.pdf.

[34] Information in this paragraph is from Environmental Defense Fund, *Flood Loss Reduction White Paper*, available at http://www.edf.org/documents/594_FloodPolicy.pdf.

[35] R.A. Pielke, Jr. and M.W. Downton, "Precipitation and Damaging Floods: Trends in the United States, 1932-1997" in *Journal of Climate* (Oct. 2000, Vol. 13, 20, pp. 3625-3637).

[36] For a discussion of this finding for weather hazards, see S.A. Changnon, "*Human Factors Explain the Increases Losses from Weather and Climate Extremes*" in Bulletin of the American Meteorological Society (Vol. 18, 3, Mar. 2000, pp. 437-442).

[37] E. B. Abbott, Floods, "Flood Insurance, Litigation, Politics—and Catastrophe: The National Flood Insurance Program," *Sea Grant Law and Policy Journal* (Vol. 1, 1, Jun. 2008, pp. 129-155).

[38] For more information on federal coastal zone management efforts, see CRS Report RL34339, *Coastal Zone Management: Background and Reauthorization Issues*, by Harold F. Upton. In the early 1970s, Congress also considered general national land use planning legislation to foster state (and local) planning capacity and coordination; bills were reported by Senate committees in 1970 and 1972 and passed the Senate in 1972 (S. 632 in the 92nd Congress), but were not enacted. Many in Congress concluded that the challenges that national land use planning legislation was intended to address were most concentrated in coastal areas and needed immediate attention. The result was the enactment of the Coastal Zone Management Act with a promise by some congressional leaders to continue to pursue national land use legislation. These leaders stated that they intended to fold coastal management into this more encompassing legislation at a later date. Comprehensive land use planning legislation was never enacted, and Congress has not ventured beyond the CZMA with this approach to resource planning and management.

[39] Some changes come about by agency action without congressional direction. For example, the Corps established the National Flood Risk Management Program in May 2006 for the purpose of integrating its flood risk management programs and activities, both internally and with counterpart activities of the Department of Homeland Security, FEMA, other Federal agencies, state organizations and regional and local agencies.

[40] The report is available at http://www.iwr.usace.army.mil/ncls/docs/NCLS-Recommendation-Report_012009_DRAFT.pdf.
[41] For another discussion of the challenges of recovering from a disaster, see CRS Report RL34087, *FEMA Disaster Housing and Hurricane Katrina: Overview, Analysis, and Congressional Issues*, by Francis X. McCarthy.
[42] UMRCP final report.
[43] Ibid.
[44] The Mississippi River Commission press release is available at http://www.mvd.usace.army.mil/offices/pa/releases/ 2008/RelMRC0801.pdf.
[45] Letter to Hon. James M. Inhofe, Ranking Member Committee on Environment and Public Works, from Assistant Secretary of the Army (Civil Works) John Paul Woodley, Jr., transmitting the UMRPC final report, Jan. 15, 2009.
[46] The Tennessee Valley Authority prepared for the Corps, *An Economic Evaluation of Proposed Flood Protection Plans on the Upper Mississippi River and Illinois Waterway* (Oct. 2004), available at http://www2.mvr.usace.army.mil/UMRCP/Reports.cfm. The report states "As flood risks are reduced in floodplains, the likelihood of economic activity may increase ... Portions of land previously zoned to prohibit development could become usable." (p. 2)
[47] Appendix B of the May 2006 draft of the Upper Mississippi River Comprehensive Plan May 2006, available at http://www2.mvr.usace.army.mil/UMRCP/Reports.cfm. (It is unclear whether updated appendices accompany the June 2008 UMRPC final report.) UMRCP final report stated "for the purposes of this study, it is assumed that whatever climate changes occur within the 50-year planning timeframe will have little effect on the types of vegetation, cropping patterns or flood frequencies as currently determined." (p. 51)
[48] 1994 Galloway report, pp. 67-68.
[49] UMRCP final report, p. ES-13.

In: Federal Flood Policy
Editor: James E. Rysanek

ISBN: 978-1-61324-017-5
© 2011 Nova Science Publishers, Inc.

Chapter 2

FEMA'S PRE-DISASTER MITIGATION PROGRAM: OVERVIEW AND ISSUES[*]

Francis X. McCarthy and Natalie Keegan

SUMMARY

Pre-Disaster Mitigation (PDM), as federal law and a program activity, began in 1997. Congress established a pilot program, which FEMA named "Project Impact," to test the concept of investing prior to disasters to reduce the vulnerability of communities to future disasters. P.L. 106-390, the Disaster Mitigation Act of 2000, authorized the PDM program in law as Section 203 of the Robert T. Stafford Disaster Relief and Emergency Assistance Act.

From its beginnings as "Project Impact" to its current state, the PDM program has grown in its level of appropriated resources and the scope of participation nationwide. Along with that growth have come issues for Congressional consideration, including the approach for awarding grant funds, the eligibility of certain applicants, the eligibility of certain projects, the degree of commitment by state and local governments, and related questions.

Authorization for the PDM program expires on September 30, 2010. In the 111th Congress, Representative Oberstar and other sponsors introduced H.R. 1746 to re-authorize the program for an additional three years at $250 million per year and to remove the sunset provision. The bill would also

---

[*] This is an edited, reformatted and augmented version of a Congressional Research Services publication, dated February 18, 2010.

increase the minimum amount each state can receive from $500,000 to $575,000.

H.R. 1746 includes provisions that have been part of appropriations statutes that award funds both through a formula (with, as noted, a minimum amount available per state) as well as a competitive process for the majority of the funds. H.R. 1746 was approved by the Transportation and Infrastructure Committee on April 2 and was approved by the House under suspension of the rules on April 27, 2009. The language of H.R. 1746 was also incorporated into H.R. 3377, the Disaster Response, Recovery and Mitigation Enhancement Act of 2009, which was reported out of the Transportation and Infrastructure Committee on November 5, 2009. It is notable that the Administration's budget for FY2010 requested that the competitive process be dropped in favor of a risk-based assessment by FEMA. The Administration's budget for FY2011 does not contain any reference to a risk-based assessment by FEMA. However, Congress may wish to hear more regarding the risk-based allocation formula before enacting the authorizing legislation for the coming years.

In another major development in FY2008, Congress directed 95 grants to 28 states, which totaled close to 44% of all PDM funds (P.L. 110-161, Consolidated Appropriations Act, 2008). These were the first such earmarks for the PDM program. While some of the projects meet PDM eligibility standards, others may be considered emergency preparedness projects which are not eligible for grants, as defined by the Stafford Act and the PDM guidance. For FY2009, the Congress directed 51 grants to 27 states at a program cost of just under $25 million. The FY2010 DHS Appropriations measure had a funding level of $100 million with just less than $25 million for Congressionally directed projects. The listing of directed grants for the last two fiscal years provides information on jurisdictions but does not have details on the types of projects involved. In consideration of the FY2010 appropriations, amendments were offered in the House and Senate to curtail the earmarking of PDM funds.

OVERVIEW OF PRE-DISASTER MITIGATION

Program Purposes

The purpose of the original pre-disaster hazard mitigation pilot program, known as "Project Impact," as well as the successor Pre-Disaster Mitigation (PDM) program, has been to implement hazard reduction measures prior to a disaster event. Those measures are similar to those actions taken following a disaster under the authority of the Section 404 Hazard Mitigation Grant Program

(HMGP).[1] The range of eligible projects might include retrofitting public buildings against hurricane-force winds or seismic damage, acquiring and relocating properties out of a flood plain, elevating structures in a flood plain, flood-proofing public buildings, vegetation management to mitigate against wildfires, or constructing or converting public spaces into "safe rooms" in tornado-prone areas.

While there would appear to be general agreement among analysts and practitioners on successful mitigation measures, there is continuing debate on where the line is drawn between preparedness for response to the next disaster and mitigation measures to lessen its impact. A common distinction frequently drawn is between structural and non-structural mitigation. Structural mitigation is the building of levees to protect communities from flooding, such as those constructed by the U.S. Army Corps of Engineers. A non-structural mitigation project would be to establish new land use patterns, and possibly remove structures from a flood plain that has repeatedly experienced flood damage. The essential difference is that the structural projects tend to construct barriers to protect communities, while non-structural projects remove structures and citizens from harm's way. The removal of homes from a flood plain is an example of the type of project eligible under HMGP and PDM.

Context and Trends

When Congress first appropriated funds in FY1997 for mitigation activities before disasters occur, FEMA established the pilot program and called it "Project Impact." The communities participating in the initial pilot program were selected by FEMA based on factors such as their experience with natural disasters, the ongoing risk the community faced, and the degree of collaboration among local, county and state officials. Project Impact placed most of its emphasis on community efforts to mitigate those hazards that made the community vulnerable to future damage.

This emphasis on community-based efforts included the required commitment of the local governments, non-governmental organizations, the local business community, as well as the development of an educational component for community awareness. This approach grew out of experience which demonstrated the necessity of community "buy-in" and active involvement with mitigation activities.

The study of elite attitudes and opinions with respect to disaster mitigation policies demonstrates the relatively low priority placed on natural hazards as political issues in local communities and even at the state level. It further demonstrates the relative unpopularity of nonstructural mitigation measures as compared to structural solutions to disaster problems or to traditional relief and rehabilitation policies.[2]

While noting the reported reticence toward nonstructural mitigation, some in the field were also turning a critical eye toward structural mitigation as a panacea for the risks posed by natural hazards. One observer spoke to the gaps in the policy area as follows:

Structural mitigations, for example, encourage people to move into hazardous areas. Post-disaster relief tends to socialize risks, lets people be insensitive to hazard risk when they build structures, and so forth. The current emphasis on nonstructural or land use approaches reflects a concern that previous policy emphases may well have increased, rather than decreased, the level of population at risk from hazards.[3]

The concept of disaster mitigation had been favorably discussed for several decades among some in the emergency management field. But absent serious disaster damage during most of the 1980s, it was difficult to advance the concept. As one observer explained:

With the comparative absence of major disasters during the Reagan years, priorities shifted and commitment to proactive measures requiring time and money waned. But in the early 1990's, that attitude dramatically changed. Massive losses between 1989 and 1993 from five major hurricanes, earthquakes, and river floods resulted in mitigation making more sense to more people than at any time previously.[4]

As noted above, the relative quiescence of the Reagan years from an emergency management perspective was followed by years with disasters of great scale in both human costs and financial damages. The disasters included Hurricane Hugo (1989); the Loma Prieta earthquake (1989); Hurricane Andrew (1992); the 1993 Midwest floods; the Northridge, California earthquake (1994); and Hurricanes Fran and Floyd (1996 and 1999) along the eastern coast of the nation. The confluence of these events helped to support those in favor of proactive work to lessen the impact of disasters, but little organized research had been done up to that point to demonstrate the benefits of pre-disaster mitigation. Without such studies (later mandated by the Disaster Mitigation Act of 2000 -

DMA2K[5]), Congress approached the PDM concept cautiously and provided funding at lower levels until the benefits of such a program were proven.

PDM LEGISLATIVE AND APPROPRIATIONS HISTORY

Pre-disaster hazard mitigation activities were initially funded through a pilot program first provided for in the conference report that accompanied the 1997 appropriations legislation. The pertinent report language follows:

> The conferees agree to up to $2,000,000 for FEMA's participation in appropriate pre-disaster mitigation efforts. The conferees agree with FEMA's Director that mitigation activities can ultimately save significant sums from post-disaster clean-up and response actions and that the Agency should be taking an increasingly active role in developing and participating in pre-disaster mitigation programs. Such programs range in scope from the development and/or funding of mitigation plans for communities to participation with industries, insurers, building code officials, government agencies, engineers, researchers and others in developing systems and facilities to test structures in disaster-like circumstances. The conferees understand that these activities will require an infusion of considerable up-front financial support as well as the possible movement over time of disaster relief funds to pre-disaster programs, and the Agency is expected to use up to the $2,000,000 provided herein in an appropriate manner to begin the process of movement toward a meaningful pre-disaster mitigation program. Expenditure of these funds may not, however, be made until submission to the Committees on Appropriations of an appropriate pre-disaster mitigation spending plan.[6]

Subsequent appropriations measures for fiscal years 1998, 1999, 2000, and 2001 provided $30 million for 1998 and $25 million per year for the next three years.[7] Following four years of funding through appropriations statutes, Congress authorized the program from 2000 to 2003 in the Disaster Mitigation Act of 2000 (DMA2K) which placed the PDM program in the Robert T. Stafford Disaster Relief and Emergency Assistance Act as Section 203.[8]

Originally, in its FY2003 and FY2004 budget requests, the Bush Administration proposed consolidating all mitigation funds in the PDM program. "Adoption of this proposal would have terminated funding provided through the Hazard Mitigation Grant Program after a major disaster is declared."[9] Congress did not wish to entirely eliminate the post-disaster mitigation help but did devote more resources to the pre-disaster program. In order to shift the resource balance between post-disaster mitigation and pre-disaster mitigation, Congress reduced the HMGP amount in the Stafford Act for post-disaster work from 15% of the total amount spent on the disaster (less administrative costs) to 7.5%.[10] While the post-disaster mitigation pot would shrink, the PDM program would grow. However, this shifting of resources would be short lived.

Table 1. History of Pre-Disaster Mitigation (PDM) Appropriations, FY1997 to FY2008

Fiscal Year	Program	Amount Requested (in millions)	Appropriations (in millions)
1997	Project Impact	N/A	$2 EMPA account[a]
1998	Project Impact	$50	$30 EMPA account
1999	Project Impact	$50	$25 EMPA account
2000	Project Impact	$30	$25 EMPA account
2001	Project Impact	$30	$25 EMPA account
2002	Project Impact	$0	$25 EMPA account
2003	PDM	$300	$150 PDM Fund established[b]
2004	PDM	$300	$150 PDM Fund
2005	PDM	$150	$100 PDM Fundc
2006	PDM	$150	$50 PDM Fund
2007	PDM	$100	$100 PDM Fund
2008	PDM	$75	$114 PDM Fund
Fiscal Year	Program	Amount Requested (in millions)	Appropriations (in millions)
2009	PDM	$75	$90 PDM Fund
2010	PDM	$150	Pending

Source: FEMA, Mitigation Directorate, January 2010.
a EMPA is the Emergency Management and Planning Assistance (EMPA) account, which is FEMA's general administrative account.
b The separate PDM account creates a separate line item for PDM for the first time in the FEMA budget.
c For the first time in legislative language P.L. 108-334 directed that the PDM funds "shall be awarded on a competitive basis."

Over its dozen year history, the funding levels for PDM have risen and fallen and risen again. During this time the program also was given its own separate line item account within the DHS/FEMA budget. The changes in the funding levels represented differing approaches not only to PDM but to the mitigation concept as a whole. The 111th Congress has introduced legislation (H.R. 1746 and H.R. 3377) that would stabilize funding by authorizing the appropriation of $250 million each fiscal year for 2010, 2011, and 2012.[11] The proposed legislation would also remove the sunset provision, though authorization for funding after FY2012 is undetermined.

The original "Project Impact," the first PDM program, was closely identified with then FEMA Director James Lee Witt. Witt was appointed by President Clinton in 1993 and gained a high profile in the course of leading FEMA's disaster response and recovery efforts. Witt described "Project Impact" as "a program designed to break the damage-repair, damage-repair cycle and instead help communities become disaster resistant."[12]

While the initial funding amounts were relatively small for a national program, Project Impact was generally considered a success. One author observed, for example, that "the money was said to have worked wonders."[13] However, some observers maintained that if funding were provided through a competitive process the criteria could recognize areas with the greatest risk and where mitigation measures could produce the most beneficial results, rather than areas that may have experienced random disasters but did not necessarily face as grave an ongoing threat.

Early in the George W. Bush Administration, "Project Impact" was eliminated from the FY2002 budget on the same day that the Mayor of Seattle was praising the program for preventing further damage due to the Nisqually earthquake.[14]

When the PDM authorizing legislation (DMA2K) was passed, Congress addressed some of the same themes used in "Project Impact" but placed the responsibility on the Governor of each state to suggest up to five communities to be considered for pre-disaster mitigation assistance.[15] While the Governor nominated potential grantees, FEMA made the final selections. In addition, under the statute, FEMA had the discretion under "extraordinary circumstances" to award a grant to a local government that had not been recommended by a Governor.[16]

In 2002 FEMA had decided to re-brand "Project Impact" the Pre-Disaster Mitigation (PDM) program. While this title conformed to the legislative language it also was intended to send another message as then FEMA Director Joe M. Allbaugh explained:

I want to take the "concept" of Project Impact and fold it in to the program of mitigation. Project Impact is not mitigation. It is an initiative to get "consumer buy-in." In many communities it became the catch-phrase to get local leaders together to look at ways to do mitigation.[17]

For FY2003 and FY2004, Congress increased funding for pre-disaster mitigation to $150 million from the previous $25 million level. Also, Congress had inserted legislative language in the FY2003 Appropriations Act, which became law on February 20, 2003, stating that PDM funds "shall be awarded on a competitive basis."[18] FEMA conformed to the direction from Congress and made part of PDM a competitive grant program thereafter.[19]

While the authorization of PDM in FY2000 had recognized, at a minimum, the potential benefit of mitigation prior to disaster events, the substantial funding increase beginning in FY2003 was one component of a different overall approach. This new approach was targeted not only to pre-disaster mitigation but to mitigation in general. It represented a shift in thinking regarding the most appropriate time to devote resources to mitigation in disaster-prone communities.

Some had suggested that the Hazard Mitigation Grant Program (HMGP) in the Stafford Act (Section 404), which provides funding to a state following a major disaster to mitigate future disaster damage, was taking the wrong approach, or, more precisely, was in the wrong sequence. Since the funds arrive after the disaster event, and are only available to states that have suffered the impact of a disaster, they cannot be targeted at areas that might have a greater risk of a more costly disaster that has not yet occurred. Pre-disaster mitigation, they argued, would be more effective.

However, others contended that only communities that have had recent disaster experience have the immediate incentive, in the form of a community commitment borne of experience, to take the steps necessary to reduce the risk of future disasters. As one writer in the field has noted, it is imperative to garner community support around a specific action:

> This is especially true when those mitigation measures involve cranking up the machinery of government, which, some contend, is especially prone to inertia.... Mitigation measures are also most effective when they have the broad support from the greatest number of people across a broad section of the community.[20]

MITIGATION FUNDING AND STUDIES

Following Hurricane Katrina, Congress chose to reinstate the HMGP to its previous level of 15% for the majority of disasters and established a new graduated scale for larger events.[21] With that change, smaller amounts were requested and appropriated on an annual basis for the PDM program. In FY2006, the appropriated amount was $50 million. However, since then Congress has appropriated larger sums for the PDM program, equal to or above requested levels.

These increases coincide with studies released in 2005 and 2007, each of which pointed to savings of $3 to $4 for each $1 spent on mitigation.[22] The findings of these studies were important to the PDM program:

> provide independent evidence to support what nearly every member of the hazards community knows anecdotally—generally, FEMA mitigation grants are highly costeffective.[23]

One study, *Natural Hazard Mitigation Saves: An Independent Study to Assess the Future Savings from Mitigation Activities*, in accordance with the directive from P.L. 106-390, was completed by the Multi-Hazard Mitigation Council (MHMC). The MHMC study defined a broad number of benefits that reached into not only direct FEMA disaster costs but also assessed corollary and indirect savings from mitigation at the local level and within the business sector with an impact, or "ripple effect" on the surrounding communities. The study weighed damages that were not always previously considered when calculating savings, such as business interruption and environmental costs. The study, released in 2005 before the hurricane season, provided a foundation for mitigation that was previously based on anecdote and conjecture. The MHMC study listed areas of savings within communities from mitigation and also focused on the long-term beneficial effects that mitigation activities would have on the federal treasury on an annual basis.[24]

Building on the MHMC study, in 2007 the Congressional Budget Office issued its report on pre-disaster mitigation cost savings. While using slightly different assumptions and cognizant of federal spending time lines, that report also noted a proportional savings derived from the PDM program. The CBO study explained that PDM savings would likely benefit two FEMA programs.

> Any federal savings from PDM-funded mitigation projects would occur largely in FEMA's disaster relief programs (which are funded from discretionary

appropriations) and in its National Flood Insurance Program (which ordinarily is not funded through the appropriation process).[25]

These findings provided a justification for increased PDM funding, which followed in FY2007.

Post-Katrina Funding—Competitive and Formula Grants

During FY2007 Congress increased PDM funding to $100 million, raised that amount to $114 million in FY2008, and in FY2009 reduced that amount to $90 million. In recognition of the larger appropriated levels, Congress directed FEMA to implement the state minimum of $500,000 specified in the Stafford Act[26] for eligible projects.[27] This formula, in effect, made PDM both a competitive and a formula-driven program. The implementation of the state minimum also served to retain interest in mitigation for states that may not have been competitive, nor experienced recent disasters.

The overall change in formula created a new kind of hybrid program, in which grants would continue to be awarded through a competitive process and also through guaranteed formula amounts for each state ($500,000) with eligible projects or plans. For example, from a total program amount of $100 million, up to $25 million is in the formula pool and the remaining $75 million is available for the competitive grants.

The Congressionally directed spending for FY2008 PDM grants, the first earmarks for the PDM program, accounted for over $50 million or 44% of the funding. After factoring in state minimums, the available amount for open competitive grants was reduced from three quarters to just over a third of the total funds. The directed grants for FY2009 total $25 million, or just over 27% of the appropriation. Taken together, the earmarks combined with the state minimums could total $50 million or 55% of the total appropriated program funds. In reaction to this trend, amendments were offered in each chamber, during consideration of the FY2010 appropriations bill, to curtail the earmarks. The Senate amendment would have eliminated the earmarks from the FY2010 appropriations.[28]

In addition to Congressionally directed spending, FEMA has established a program rule that governs the size of respective grants in FY2009.

> States and territories that submitted less than $500,000 in applications received the amount requested, provided those applications are determined to be

eligible. The maximum PDM award for any one State shall not exceed $17 million. There is a $1 million cap on the federal share available for plans and a single federal share cap of $3 million for projects.[29]

The Bush Administration requested $75 million for FY2009. Congress funded the program at the $90 million level. The budget justification submitted to Congress for the FY2009 budget noted the $39 million reduction from the FY2008 level did not offer any comment or explanation for the change. Some have suggested that the seeming carryover amount between FY2007 and FY2008 of more than $65 million may have contributed to the conclusion that additional funding was not needed. FEMA has noted that since PDM funds are no-year funds with a great amount of state and local participation in the process, the lag time on the expenditure of funds is a practical and inevitable part of program administration. FEMA has also emphasized that funds being carried over are funds dedicated to projects that have been selected and are only awaiting final clearance.

Grant Applications and Categories

Given the authorizing language that requested that each Governor submit "not fewer than five local governments to receive assistance under this section"[30] it is not surprising that the program would have a large number of grant awards (a total of 149 grants were awarded for FY2008 and 443 applications were received for FY2009). The total number of grant awards is amplified by the significant number of planning grants. In FY2008, planning grants accounted for 79 percent of the awarded grants. These are usually awards for much smaller amounts than project applications, and planning grant awards are distributed to many more communities. The interest in planning may derive from the fact that a mitigation plan is a prerequisite for receiving both PDM and HMGP funding.

Grants have been awarded for a variety of hazards being addressed by states and communities. The Government Accountability Office (GAO) reviewed the FY2003 projects and found that more than half of the projects identified flooding as the primary hazard being mitigated by the grants. That same review found that 12% of the grants were based on hurricane projects, just under 7% sought to mitigate the effects of an earthquake, and 4% listed tornadoes as the primary hazards to be addressed.[31]

The PDM projects funded at the direction of Congress for FY2008 also sought to accomplish a variety of purposes. Some appear to be traditional PDM projects such as the acquisition and relocation of properties and wildfire mitigation activities. However, other projects listed among the earmarks appear to be for purposes listed as ineligible in the PDM program guidance materials. Examples of those projects include funding for equipment, fire suppression activities, dams, and emergency alert and notification systems. These projects reflect the preparedness vs. mitigation debate that, as the "Program Purposes" and "Funding Criteria" sections of this chapter explain, has been with the PDM program since its inception.

ISSUES FOR CONGRESSIONAL CONSIDERATION

As Congress considers re-authorization of the PDM program there are several issues that have emerged as points of discussion. These issue include the pace of funding distribution, the best methods for funding awards, the priority uses for PDM funds, the amount of resources devoted to the program, and the length of authorization for the program. Also, new initiatives emerged from the 2010 budget and authorizing legislation that suggested new directions for the PDM program.[32]

The Pace and Breadth of PDM Funding Distribution

As previously noted, in FY2008 the PDM program was earmarked for the first time.[33] The PDM program was earmarked again in the FY2009 appropriation.[34] The only previous earmarks of mitigation projects in general appeared in the FY1999 Appropriations bill that earmarked unspent and prospective HMGP funds for several projects.[35] Exact amounts of funding and the rate at which such grant funds are disbursed can be difficult to discern, but the broad geographic distribution of recipients has been a constant in the PDM program. The congressionally directed earmarks for the FY2008 and FY2009 add to that distribution across many jurisdictions.

The funds have been distributed widely, but not always rapidly. While the earmarks are new to the program, some have pointed to the lags in PDM spending, such as the carryover of funds from FY2007 to FY2008, as an explanation for the earmarks. Others have suggested that the same lag in funding, interpreted as a lack of interest in or need for the program, may have resulted in a reduced request by the Administration for FY2009 PDM funding.

Table 2. Recent Distribution of PDM Funds, FY2006 to FY2008

Agency	FY2006 Recipients	FY2007 Recipients	FY2008 Recipients	FY2009 Recipients
DHS/FEMA	67 grants in 37 states, 1 territory	249 grants in 44 states, 1 territory	144 grants in 44 states, 1 territory	137 grants in 43 states, 1 territory
DHS/FEMA	4 grants to 4 Indian Tribal Governments	6 grants to 6 Indian Tribal Governments	None	1 grant to 1 Indian Tribal Government
Congressional Direction	N/A	N/A	113 grants in 27 statesa	52 grants in 27 statesb

Source: All information for years FY2006, FY2007, 2008 and FY2009 are from FEMA, Mitigation Directorate, July 7, 2009.

a The first total for grants numbers and states for FY2008 include the projects identified in the House Appropriations Committee print of Congressional earmarks for the PDM program. The initial number of projects listed under P.L. 110-161 totaled 95 projects. The increase is based on FEMA's engagement with selected communities and developing more eligible mitigation projects.

b The initial number of projects listed under P.L. 110-329 for FY2009 totaled 52 projects. This number could expand as FEMA works with the local communities receiving the earmarked funds to determine eligible projects.

One consideration in the pace of the program is that mitigation projects can be complicated to put together since their impact may be spread across various sectors of communities and can also require local consensus and a contribution of resources. The state and local cost share is 25%.[36]

Another possible factor in the arguably slow pace is that PDM funds are available until expended. Since, under the PDM program's guidance, the funds can be used for up to three years from the date of the award some may contend there is less urgency to get funds out immediately and more time for communities to develop effective projects and plans and more time for FEMA, through a peer review process, to carefully review the submitted projects and plans.

The perception of slow distribution of PDM funds has continued in later years as evidenced in the pace of awards made. According to FEMA listings, in FY2006 when $50 million was made available, only $39 million was awarded.[37] Similarly, for FY2007 $100 million was appropriated, but only $52.3 million had been awarded, and for FY2008 awards are still pending according to totals on the FEMA website.[38]

However, FEMA staff have provided updated figures that now place total FY2007 funding distributed at $131 million for a year when $100 million was appropriated. These larger figures represent funding for planning and projects carried over from previous years.[39] This approach to batching funding was officially used by FEMA in FY2005.

> Approximately $255 million is available for competitive grants, technical assistance, and program support for the FY 2005 PDM program. As PDM funds are available until expended, this amount is comprised of Fiscal Year 2003, 2004, and 2005 funds.[40]

Also, when assessing funds not allocated to awarded grants it is helpful to understand how the unallocated program dollars are used. Some of those funds are devoted to ongoing expenses for each program year including FEMA administrative costs, technical assistance contracts to assist applicants and sub-applicants, management costs awarded to states, and other costs associated with the award amounts. FEMA also holds back a small amount of funding for "reconsideration" which allows for the review of projects and the correction of possible errors in program administration, grant selection, and the calculation of funding amounts.[41] All of these factors, from FEMA's perspective, are reasonable uses for unexpended funds. FEMA has recently issued a chart that identifies the broad uses of program funds.

Table 3. PDM Funds (in millions)

Program	Total Appropriations[a]	Admin. Program Support/Technical Assistance[b]	Total Obligated[c]	Applicant Management Costs	Remaining Funds for Grants
PDM	$664,000,000	$66,400,000	$478,348,302	$19,272,526	$99,979,172

Source: FEMA Mitigation Directorate, March, 2009.
Notes: Click here and type the notes, or delete this paragraph
a Totals from program inception through FY2008.
b PDM – 3% admin. 7% program support and technical assistance.
c Total obligated does not include administrative, management and technical costs.

The reserved funds and other costs can be problematic, however, when they are not identified in program lists of award amounts and are estimated as a percentage of annual program costs. Similarly, FEMA's approach to batching together several years of project funding may be a reasonable approach to multi-year projects, but is not explained in the fiscal year totals currently available to the public. These kinds of issues in how funding awards and other spending are reported can present problems to Congress in assessing the program as a whole.

Terrorism and Pre-Disaster Mitigation

Some have questioned whether the PDM funding should be available to mitigate the effects of terrorist events. The response of some PDM advocates is one that applies not only to purpose but particularly to the overall balance of resources between mitigation and preparedness programs. Some participants in this debate have noted that while some projects may arguably be considered preparedness or mitigation, there is little similarity between funding amounts available for those two purposes, nor for the programs addressing terrorism.

While funding for the PDM program previously exceeded $100 million, the amounts for preparedness efforts for all-hazards, including terrorism, under DHS/FEMA grants has totaled in the billions at DHS/FEMA in recent years. Among those preparedness programs at FEMA, several of the grant programs permit the purchase of equipment such as warning systems and other preparedness projects sometimes requested under the PDM program.[42] Perhaps most importantly, the authorizing language for the PDM program specifically makes clear that the state and local governments interested in participating in the program are expected to identify "natural disaster hazards" in areas under their jurisdiction for mitigation work.[43]

Projects and Plans

As noted earlier, grants for protecting public buildings or private residences are the awards most closely associated with PDM. Projects tend to be costly and relatively large in scale when allocated for the purposes of relocating neighborhoods, building large safe rooms, or undertaking similar expensive, structural work. However, another significant category of eligible work under the PDM program is the creation or improvement of hazard mitigation plans for a community or state. With the passage of P.L. 106-390, the Disaster Mitigation Act of 2000 (DMA2K), planning took on much greater significance. In addition to authorizing PDM, DMA2K also authorized the requirement for mitigation planning and authorized increasing the share of HMGP grants from 15% to 20% of total disaster spending for states with an "enhanced mitigation plan."[44] The complementary nature of the Stafford Act hazard mitigation authorities is arguably evident when states use PDM funds to develop the "enhanced plans" that, when approved, result in higher levels of HMGP funding.

Such planning grants are a major component of the PDM program. In FY2006 the planning grants comprised 47% of total grants selected for further review; in FY2007 59% of such grants selected for further review were for planning efforts; and, in FY2008, of the 149 proposed projects, 117 were identified as planning grants.[45] However, the actual funding amounts for planning are relatively low. During FY2006, projects selected for further review projected grant spending of $42.8 million while planning grants selected for further review totaled $3.9 million out of a total of $50 million.

Similarly, in FY2007, the large majority of planning grants (135 of the grants selected for further review) totaled only $16.5 million while project grants selected for further review (75 grants) were awarded $67.1 million out of $100 million available for awards.[46] Given the nature of project grants and the large undertakings they represent (such as property acquisitions and similar commitments), they are far more expensive than planning grants.

Table 4. Planning Grants and Project Grants

Fiscal Year	Planning Grants Selected	Project Grants Selected	PDM Program Funding (millions)	Planning Grants in Dollars (millions)	Project Grants in Dollars (millions)
FY2006	47%	53%	$50	$3.9	$42.8
FY2007	59%	41%	$100	$16.5	$67.1
FY2008	79%	21%	$114	$12.3	$27.7

Source: FEMA Mitigation Directorate.

The remaining $20 million for the FY2007 awards includes awards still being made, administrative costs, technical assistance for applicants, state management costs, and funds held back for reconsideration.[47]

Resources vs. Requests

The importance of the actual amount of funds appropriated to the program is apparent when reviewing the amounts available for PDM grants alongside the amounts requested by applicants. In FY2006 and 2007, for example, the funding requested was nearly triple the amounts available. In FY2006, $50 million was available and FEMA received initial requests totaling $134 million. In FY2007,

FEMA had $100 million available for grants and received requests for $292 million.[48] Given the limit of five applications per state, it is reasonable to suggest that the amounts requested could have been even higher absent that limitation.

Funding Criteria

The authorizing legislation for PDM sets forth an array of funding criteria. The criteria focus on elements such as the nature of the hazard, the degree of commitment of and coordination by the state and local governments (including consistency with appropriate mitigation plan), and the "extent to which prioritized, cost-effective mitigation activities" can produce clear results.[49]

Along with the statutory funding criteria, FEMA, in its PDM program guidance, lists ineligible activities for PDM planning and project activities. FEMA staff noted that they have derived many of the suggested changes from the eligibility listings from the peer review panels, composed of local practitioners in the mitigation/emergency management field, that review applications each year. It is the intent of the program staff to provide more clarity on eligible activities for applicants by providing such a list.[50]

The ineligible activities list for FY2008 contains eight items related to PDM planning and 23 ineligible activities for the PDM project grants. (For the latter category, this is an increase; for FY2007, the number of ineligible activities was 16).[51] The list broadly supports compliance with practices such as environmental and historic preservation and the Coastal Barrier Resources Act (CBRA). But other excluded items (such as the construction of levees or flood mapping) are arguably seeking to ensure that PDM planning or project funds do not duplicate similar efforts funded by other programs.

However, some observers argue that the FEMA interpretation of eligible PDM projects has grown overly restrictive, particularly with regard to equipment purchases to address different hazards. For example, some observers believe that the purchase of warning or alert notification systems should be an eligible expense for PDM. (It should be noted that warning systems and other "gray areas" can be funded through the HMGP program's 5% initiative that was put in place a dozen years ago. This was established to allow some flexibility for actions that may or may not meet cost-effectiveness criteria).[52] Others suggest that the purchase of generators under the PDM program should be eligible beyond the standards for such purpose in the program guidance.[53] The arguments over individual categories and projects are symbolic of the overarching effort to differentiate the concepts of preparedness and mitigation.

Project Eligibility

There are a number of project activities that are ineligible under FEMA's program guidance for the PDM program. Some of the ineligible activities include costs of maintenance to structures (e.g., levees and dams); the purchase of generators for facilities that are not a part of a larger mitigation project; and the broadest category – projects for which benefits "are available from another source for the same purpose."[54]

A particular example at the crux of this debate concern warning systems. Many communities have sought to use PDM funds to purchase warning systems such as sirens to protect their citizens against sudden disasters. FEMA considers such alert notification systems as eligible under disaster preparedness grants but not under the PDM program. Similarly, FEMA has previously determined that the purchase of stand-alone generators is a preparedness effort to address the likely results of a disaster rather than mitigating its effect. One exception is the purchase of generators that will power a mitigation effort. For example, a generator providing power to activate hurricane storm shutters would be eligible. Generators that provide power for critical public facilities may also be eligible.

For FY2008, some of the congressionally earmarked projects for PDM include some of the activities listed as ineligible in FEMA's program guidance such as fire suppression activities and the purchase or enhancement of emergency alert and notification systems. Such designations do not involve differences over the location of grants but their purposes. (The FY2009 listing of earmarks did not list the type of project or purpose.) Congress may wish to express its disagreement with FEMA's guidance or it may direct FEMA to adhere to the PDM program's current eligibility criteria when making PDM grant awards.

Length of Authorization

The PDM program has been re-authorized previously in six different pieces of legislation, initially for three years, then two one-year re-authorizations through appropriations bills, and then another three-year authorization from 2005 to 2008 followed by a one-year authorization for FY2009 and a one-year authorization through an appropriations bill for FY2010.[55] The 111th Congress is considering legislation that would provide a three-year authorization through FY2012.[56] The original sunset date of P.L. 106-390 (December 31, 2003) was intended to provide time for more information to be gathered on the efficacy of pre-disaster mitigation. Some of that has been presented in both the Multi Hazard Mitigation

Council Report as well as the report by the Congressional Budget Office. The recurrent sunset date, however, has set the PDM program apart from the rest of the Stafford Act which is a free-standing, no-year authorization. If the initial questions concerning the efficacy of the program are resolved, Congress might authorize the PDM program, like the rest of the Stafford Act, without a sunset date..

On the other hand, it can be argued that some of the Stafford Act provisions are so vital to emergency situations (e.g., debris removal, temporary sheltering and lodging) that not having to seek re-authorization on a regular basis is a practical and effective approach to the disaster response and recovery aspects of the statute. Conversely, since the PDM program is a grant program not funded from the Disaster Relief Fund (DRF), some might contend, having a three to four year re-authorization cycle provides incentives to all participants to refine and improve the program in anticipation of Congressional oversight. Also, through annual appropriations Congress can actively evaluate the PDM program accomplishments.

Methods of Awarding PDM Funds

When the pilot program, Project Impact, was initiated in 1997 an emphasis was placed on the communities' disaster history, the involvement of community-based organizations in mitigation work, the participation of the local business community and the commitment of the state and local governments. There was some concern at the time on the part of state emergency management officials that they were not sufficiently involved during the project selection process. The switch to a competitive process in PDM reflected some of those factors that Project Impact employed, but also placed greater emphasis, in light of statutory language, on cost-benefit ratios. Also, since funding for planning was made eligible, the program opened up to many communities that desired an improved mitigation plan. [57]

For the overall awards process, Congress generally has come to direct the PDM program in annual appropriations law rather than through Congressional hearings specifically on the PDM program and resulting authorizing legislation.

State emergency managers have stated their position that a competitive process may tend to limit smaller states' ability to access a program like PDM. Echoing the tenets of federalism, they would like funds made available to each state and decisions made at the state and local level concerning the hazards that pose the most significant threats and the areas that could benefit most from PDM funding. As one state emergency management director, speaking on behalf of the National Emergency Management Association (NEMA), testified:

> Attempting to prioritize limited predisaster mitigation funding on the national level is counterproductive to the establishment of state and local planning, therefore NEMA supports the distribution of predisaster mitigation funds by a base plus population formula rather than by competitive grants. The competitive system as it is presently funded creates more losers than winners: in an enterprise that seeks to encourage communities to engage to protect themselves, it seems counterproductive to pit good programs against good programs when the objective is that predisaster mitigation programs be undertaken.[58]

Since 2007, in addition to the competitive process, PDM administrators have implemented a $500,000 minimum per state for eligible projects or plans. [59] Given the amount of appropriations, this minimum amount means that close to 25% of funds may now be awarded outside of the competitive process. Depending on the number of Congressionally directed projects, that percentage may be much larger.

Congress may consider examining the PDM program to return to its initial form of award selection by Governors and the President, or establish a strictly competitive grant process. A third option is the present configuration of a hybrid program that is competitive but with some flexibility for awards for every state. Congress can also consider if it wishes to continue with congressionally-directed spending that was initiated in FY2008 and, if so, at what level since it accounted for nearly half of all spending. H.R. 1746 and H.R. 3377, both of which reauthorize PDM, set the amount for state minimums for PDM funding at $575,000 per state. The Administration's budget request for FY2010 presented an additional option for the management of the program. However, the FY2011 budget does not mention that risk-based option.

Allocations vs. Competition

The FY2010 budget called for a new approach to the distribution of funds. Given the state minimum awards of $500,000 each and the Congressional earmarks, the remaining total funds to be distributed on a competitive basis has diminished to a much smaller amount. In reaction to this trend, the administration suggested jettisoning the competitive formula (which requires a large peer group panel and a lengthy judging process) in favor of a risk-based allocation formula that would simply continue the distribution to states based on FEMA's assessment of the risk. This approach would still leave discretion in the hands of the states to determine their priorities for individual projects. FEMA has done work in risk assessment, particularly its HAZUS program that estimates damage based on assorted disaster scenarios. FEMA defines HAZUS as

> a powerful risk assessment methodology for analyzing potential losses from floods, hurricane winds and earthquakes. In HAZUS-MH, current scientific and engineering knowledge is coupled with the latest geographic information systems (GIS) technology to produce estimates of hazard-related damage before, or after, a disaster occurs. [60]

FEMA will be using other inputs as well to determine its risk-based allocations. While the budget suggests a new approach to the distribution of funds, the re-authorization legislation (H.R. 1746) that passed the House in the 111th Congress has written the competitive process into the proposed law. Also, the re-authorization legislation increases the minimum amount per state to $575,000, further reducing the pool for a competitive process. In its budget submission for FY2011, the Administration makes no mention of the risk-based proposal to supplant the competitive process.

Previously, the direction toward a competitive process had been contained in annual appropriations measures. This presents a question for Congress whether to accept the Administration's new initiative or continue with the competitive approach. One early indication that the new approach to risk-based allocations may not be adopted immediately was the commentary in the House Homeland Security Appropriations Report:

> As part of the budget, FEMA requested to drastically change the distribution methodology used for awarding PDM grants. However, the agency was unable to adequately articulate to the committee the ramifications or benefits of their new approach and signaled that the proposal was still being developed.[61]

The House Appropriations Subcommittee for Homeland Security said it would not approve of the change. The legislative process is not yet completed on the appropriation. But the report also noted that H.R. 1746, the PDM authorization bill which had passed the House by a comfortable margin, did not include this approach and instead seeks to place the competitive process in law. H.R. 3377, which passed the authorizing committee, also does not include the risk-based approach. Since neither piece of legislation has been completed, additional dialogue on this issue is likely.

A Different Approach to Mitigation

An entirely different approach would be to make a structural change in program delivery. Under this proposal, the PDM program and the HMGP program would move from FEMA to a newly created Federal Mitigation and Recovery Authority.

In the aftermath of Katrina there has been criticism of FEMA's uncertain role in long-term recovery as opposed to its initial role in delivering emergency response programs such as temporary housing. (The latter also drew criticism, but FEMA's authority and responsibility was not in question.) Some have suggested that a separate authority/organization with expertise in the rebuilding cycle could be partnered with mitigation programs. In this way, two important phases—building back safer while also making communities more resilient—could receive separate but complementary attention. PDM requires planning and community-wide participation, as does recovery. The roles FEMA is expected to assume are diverse and require very different skills. Some experts have noted the differing roles may not be complementary.

> However, it is not clear to us that institutional arrangements that are appropriate for implementing emergency measures after a disaster has occurred (crisis response) are also the appropriate institutional arrangements for long-term forward planning of mitigation measures before a disaster has occurred (given the three levels of government with jurisdictional mandates in this context), which in turn may not be appropriate for planning the long-term recovery of devastated regions.[62]

Direct Application for PDM Grants by Eligible Non-Profits

Currently, non-profits or non-governmental organizations (NGOs) with eligible projects must submit their applications through their local government. This process assures knowledge and approval by local authorities. Such an approach can also combine the perspective of the nonprofit with the interest of the community as a whole. The current system arguably is a reasonable construct for communication and cooperation at the local level. However, it also means that the local government officials must move promptly, and make a submission in accordance with program rules, for a project that may not be a priority or spark administrative interest. When, in particular circumstances, this could be a burden on a local government with limited resources, the plans or projects could also be submitted through a state government as well.

FEMA's Pre-Disaster Mitigation Program: Overview and Issues 73

In order to improve the efficiency of the competitive process, it may be possible to permit NGOs to submit their project directly to FEMA. However this should be done with the requirement that the application has been, at a minimum, shared with the local government at the same time so that they may be not only aware of the project, but in agreement that the project comports with local mitigation planning. This approach was suggested in testimony by an official of an association of local government emergency managers.[63] This type of approach would also give the local or state officials the opportunity to comment on the project proposal.

Others have suggested that an NGO application for a PDM grant must be a part of the state or local hazard mitigation plan. Current FEMA guidance already states that requirement.[64] Since FEMA has placed a priority on project compliance with the plans, the instructions for NGOs that currently are a part of the program guidance could be added to statutory language. The legislative criteria for PDM only require that projects submitted by a state or local government be "consistent with the mitigation plan."[65]

Upgraded Codes and Zoning

In a hearing on the re-authorization of the PDM program, Subcommittee Chair Eleanor Holmes Norton queried panelists on evaluating the status and quality of local codes and zoning as part of the assessment of PDM grant proposals.[66] It could be argued that appropriate codes would best reflect the "degree of commitment by a state or local government" that the Stafford Act lists as a consideration.[67] While Representative Norton did not endorse that approach she was interested in hearing from panelists representing state and local officials. Panelist Jim Mullen of Washington state noted the difficult and lengthy process in changing a code. Other experts have pointed out the opposition that such proposed changes can generate within a community.

> Developers, builders, and other economic interests, including individual property owners, often oppose the adoption of strict land-use regulations and building standards and too often successfully prevent their adoptions. They argue that such regulations will increase the cost of building, reduce the value of property, limit the prerogatives of property owners in terms of what they can and cannot do with their property, and make it more difficult to sell the property to others. In large measure, their arguments are valid. The question, however, is whether those concerns outweigh the potential costs of not mitigating disasters.[68]

Local codes and zoning can arguably be considered the strongest commitment to mitigation that can be made by a governmental entity. That approach, the insistence on strong local codes, has been a part of the National Flood Insurance Program (NFIP) since its inception. NFIP regulations stipulate a criterion for participation in the flood insurance program.

> the adequacy of a community's flood plain management regulations. These local regulations must be legally enforceable, applied uniformly throughout the community to all privately and publicly owned land within flood-prone, mudslide (i.e. mud flow) or flood-related erosion areas, and the community must provide that regulations take precedence over any less restrictive conflicting local laws, ordinances, or codes.[69]

Shifting more of the PDM program to a code or zoning threshold could challenge communities to a greater mitigation commitment than required under current program criteria. As one observer has noted, a dominant federal role may appear logical in the context of overall disaster spending and in its purpose to save lives and protect property. However, the perceived federal leadership and funding also may come at a price beyond the budgetary implications.

> The perception of federal benevolence discourages responsible hazard mitigation among nonfederal interests, thus contributing to the potential for greater losses in future disasters. Shirking responsibility for hazard mitigation among states and local governments may take two forms: (1) unwillingness to expend their own funds for disaster planning and hazard mitigation and (2) avoidance of the political and fiscal burdens of regulating land use in areas subject to natural hazards.[70]

While strong and effective codes may reduce the impact of hazards, local officials, it may be argued, are weighing other considerations regarding economic growth for the community, which in turn contribute to the support of many other local governmental obligations. Additionally, the PDM program is voluntary. Communities participating in the program are taking the initiative to protect their citizens and their property. In most cases, these communities are also paying the 25% cost share for the project or plan. Another consideration is that for a program that has been criticized for its pace of expenditures, linking such spending to the development of codes or changes in zoning laws would likely create a far more lengthy application and award process.

Congress' continuing interest in this area can be noted in legislation that seeks to link mitigation concepts with zoning. The proposed legislation seeks to "enhance existing programs providing mitigation assistance by encouraging states to adopt and actively enforce state building codes."[71] The bill links support for codes to both the Hazard Mitigation Grant Program (Section 404 of Stafford) and the PDM program.

Multiple Mitigation Programs

Another issue for Congress is consideration of the PDM program within the context of federal hazard mitigation policy as a whole. However, that whole is divided among varying approaches involving timing, targeted funding for particular hazards (notably flooding), and separate funding accounts within FEMA.

Earlier in this chapter the relationship was noted between the PDM program and the post-disaster HMGP program. In addition to those two programs, FEMA also administers the Flood Mitigation Assistance Program (FMA), which is part of the flood insurance program, the Repetitive Flood Claims Program (RFC) and the Severe Repetitive Loss Program (SRL). These five mitigation grant programs have some differences, but generally fund similar projects. The history behind the programs indicates Congressional intent to address specific problems and also provide discretion to state and local governments in the manner they choose to address specific hazards.

In discussing the overall impact of its programs, FEMA's Mitigation Directorate reported that the existing mitigation grant programs awarded more than $444.2 million to 1,050 projects and plans nationwide in 2007.[72] The majority of that funding came from the HMGP program, which receives its funding on a formula basis from the Disaster Relief Fund (DRF).[73] The other programs, such as PDM, FMA, and the repetitive loss programs, are individual accounts funded through the annual appropriations process.

The Mitigation Directorate at FEMA has taken steps to, if not totally blend the programs, make sure that the programs are complementary. A good example of this approach is that the guidance provided for grant applications stresses early on that it "does seek to integrate programs by allowing applications to be considered by other mitigation programs."[74] For the FY2009 grant award period, FEMA issued a Unified Hazard Mitigation Assistance (UHMA) guidance. [75] Congress has expressed its interest in this issue. In a report accompanying the House Appropriations bill, the Committee included the following directive.

The Committee notes that this program is one of several mitigation programs run by FEMA, including the Repetitive Flood Claims grant program, the Flood Mitigation Assistance program, the Hazard Mitigation Grant Program, and the Severe Repetitive Loss grant program. Each program has a different authorization, but all aim to mitigate losses from future disasters. The Committee directs FEMA to report to the Committee within six months of enactment of this Act on a mitigation strategy showing how each program contributes to mitigation goals.[76]

An issue for Congressional consideration is whether the programs should be combined for greater and more consistent impact. A subject for consideration is that the damage reductions accomplished by these mitigation programs are reflected in smaller payments from the DRF for future disaster events. Given that fact, an argument can be made that funding for a combined mitigation program could come from the DRF through an annual allocation rather than for separate events and separate accounts. A combined program could address all hazards as is the case with the PDM and HMGP programs.

An additional argument can be made that eventual savings from mitigation activities would accrue to not only the National Flood Insurance Program (NFIP) but also the private insurance industry as losses are reduced. For that reason, it might be argued, payments for at least one program, the FMA, should continue to come from the NFIP. This view of mitigation may also be an argument for the federal government and states to consider encouraging mitigation approaches through private insurers by insisting on the adoption and implementation of mitigation measures similar to the process the NFIP employs.

CONCLUDING OBSERVATIONS

Over the last decade, the Pre-Disaster Mitigation program has developed and grown as mitigation itself has become accepted federal policy. Adoption and expansion of mitigation as a beneficial approach for government has been bolstered by studies that demonstrated cost reductions following disasters due to earlier mitigation investments.

Appraisal of the PDM program is open to different interpretations and conclusions. While program staff at FEMA point to a program with flexibility and an appreciation of the regulatory challenges faced by communities carrying out mitigation projects, other observers see what appears to be the contrary, citing unspent funds and a perceived rigidity in program guidance that hinders the flexibility of local governments in accessing the PDM funding and in using it in a manner they choose. The majority of the program funds is spent on mitigation projects, but a portion of the funding is spent on the development and improvement of state and local mitigation plans. The remainder of funds are spent for technical and administrative assistance or held back for "reconsideration" for some awards.

In FY2008 and FY2009 Congress directed the funding of some PDM projects. The earmarks were broadly distributed as previous PDM funding has been. The congressional earmarks represented 44% and 27% of funds available for the competitive and set-aside PDM grants for 2008 and 2009, respectively. The congressionally directed grants also funded some projects that do not appear to be in accord with FEMA's program guidance.

The 111th Congress is considering the re-authorization of the PDM program. The legislation under consideration (H.R. 1746 and H.R. 3377) extends the program and also would codify in law previous program practices with some adjustments. In addition, there are broader considerations the Congress may wish to take up regarding federal mitigation policy in the future and the PDM program's role in that policy.

End Notes

[1] 42 U.S.C. 5170(c). For additional information on HMGP, see CRS Report R40471, *FEMA's Hazard Mitigation Grant Program: Overview and Issues*, by Natalie Keegan.

[2] James D. Wright and Peter H. Rossi, ed. *Social Science & Natural Hazards*, (Cambridge: Abt Books, 1981), p. 78.

[3] Ibid. p. 82.

[4] Robert E. Hinshaw, *Living with Nature's Extreme's: The Life of Gilbert Fowler White*, (Boulder: Johnson Books, 2006), p. 181.

[5] P.L. 106-390, Sec. 209, 114 Stat. 1571.

[6] U.S. Congress, Conference Committee, *Making Appropriations for the Departments of Veterans Affairs and Housing and Urban Development, and for Sundry Independent Agencies, Boards, Commissions, Corporations, and Offices for the Fiscal Year Ending September 30, 1997, and for Other Purposes,* conference report to accompany H.R. 3666 (P.L. 104-204), 104th Cong. 2nd Sess., H.Rept. 104-812 (Washington; GPO, 1996).

[7] P.L. 105-65, 111 Stat. 1376; P.L. 106-390, 112 Stat. 501; P.L. 106-74, 113 Stat. 1086; and P.L. 106-377, 114 Stat. 1441A-46.

[8] 42 U.S.C. 5133.

[9] For additional information see, archived CRS Report RL32242, *Emergency Management Funding for the Department of Homeland Security: Information and Issues for FY2005,* by Keith Bea, Shawn Reese, Wayne Morrissey, Frank Gottron, and C. Stephen Redhead, p. 30. (RL32242 is out of print but available from CRS upon request.)

[10] P.L. 108-7, Sec. 417, 117 Stat. 525.

[11] H.R. 1746 and H.R. 3377.
[12] James Lee Witt and James Morgan, *Stronger in the Broken Places*, (New York: Times Books-Henry Hot and Company, 2002), p. 42.
[13] Robert Block and Christopher Cooper, *Disaster: Hurricane Katrina and the Failure of Homeland Security*, (New York: Times Books-Henry Holt and Company, 2006), p.68.
[14] Ibid.
[15] 42 U.S.C. 5133(d).
[16] 42 U.S.C. 5133(d)(2)(B).
[17] U.S.Department of Homeland Security, Federal Emergency Management Agency, Testimony of Joe M. Allbaugh before the Senate Appropriations Committee, Subcommittee on VA, HUD and Independent Agencies, at http://www.fema.gov/about/director/allbaugh/testimony/051601.shtm.
[18] P.L. 108-7, 117 Stat. 515.
[19] U.S. Department of Homeland Security, Federal Emergency Management Agency, Fiscal Year 2003 Pre-Disaster Mitigation Program, at http://www.fema.gov/government/grant/pdm/fy2003.shtm.
[20] R.W. Greene, *Confronting Catastrophe* (ESRI Press: Redlands, California, 2002), p.15.
[21] Stafford Act, Section 404, as amended, 120 Stat. 1447. If Stafford Act funding does not reach $2 billion, the HMGP program will receive 15% of that amount. For disasters between $2 billion and $10 billion, the HMGP award is 10% of the total. If the disaster total is between $10 billion and $35.3 billion, the HMGP award is 7.5% of that amount.
[22] Multi Hazard Mitigation Council of the National Institute of Building Sciences, *Natural Hazard Mitigation Saves: An Independent Study to Assess the Future Savings from Mitigation Activities*, December 2005, at http://www.nibs.org/ MMC/mmchome.html, and *CBO Potential Cost Savings from the Pre-Disaster Mitigation Program*, Congressional Budget Office, September 2007 at http://www.cbo.gov/ftpdocs/86xx/doc8653/09-28-Disaster.pdf.
[23] "Mitigation Generates Savings of Four to One and Enhances Community Resilience," *Natural Hazards Observer*, vol.xxx, no.4 (March2006), p. 1 at http://www.colorado.edu/hazards/o/archives/2006/mar06/mar06a.html.
[24] Multi-Hazard Mitigation Council, National Institute of Building Sciences, *NATURAL HAZARD MITIGATION SAVES: An Independent Study to Assess the Future Savings from Mitigation Activities, Volume 1 - Findings, Conclusions and Recommendations*, 2005, pp. 2-6.
[25] U.S. Congressional Budget Office, *Potential Cost Savings from the Pre-Disaster Mitigation Program*, September 2007, p. 4.
[26] 42 U.S.C. 5133(f).
[27] U.S. Department of Homeland Security - Federal Emergency Management Agency, *FY2007 Pre-Disaster Mitigation Program Guidance*, p. 1, at http://www.fema.gov/library/ view Record.do?id=2095.
[28] Sen. Feingold's amendment #1402 to H.R. 2892 would have removed earmarked projects in both the PDM program and the Emergency Operations Center (EOC) program. The amendment failed on a vote of 60 to 38 on July 8, 2009. {http://www.senate.gov/galleries/pdcl/index.htm].
[29] Memo from Mike Grimm, FEMA Mitigation Directorate, May 13, 2008, available from the authors.
[30] 42 U.S.C. 5133.
[31] See Government Accountability Office, *Hurricanes Katrina And Rita: Unprecedented Challenges Exposed the Individuals and Households Program*, Washington, September 2006, http://www.gao.gov/new.items/d04727r.pdf.

[32] While the Administration's budget for FY2010 requested that the competitive process be dropped in favor of a risk-based assessment by FEMA, the Administration's budget for FY2011 does not contain any reference to a risk-based assessment by FEMA.
[33] P.L. 110-161, Consolidated Appropriations Act, 2008, Division E - Department of Homeland Security Appropriations Act, 2008 (House Appropriations Committee Print) pp. 1112-1115.
[34] P.L. 110-329, Consolidated Security, Disaster Assistance, and Continuing Appropriations Act, 2009 (House Appropriations Committee Print) pp. 685-687.
[35] P.L. 106-74, 113 Stat. 1086. This act contained earmarks of mitigation funds for California, Florida and North Carolina.
[36] 44 CFR 206.432(c).
[37] U.S. Department of Homeland Security, Federal Emergency Management Agency, *List of FY 2006 Pre-Disaster Mitigation Grant Recipients*, at http://www.fema.gov/government/grant/pdm/fy06_pdm_grant_recipients.shtm.
[38] U.S. Department of Homeland Security, Federal Emergency Management Agency, *List of FY 2007 Pre-Disaster Mitigation Grant Recipients*, at http://www.fema.gov/government/grant/pdm/fy07_pdm_grant_recipients.shtm.
[39] The latest figures provided by Mike Grimm, Deputy Director, Risk Reduction Division, FEMA Mitigation Directorate, in a memo as of May 20, 2008. The update presents a much different picture from the figures available on the public website.
[40] U.S. Department of Homeland Security, Federal Emergency Management Agency, *Fiscal Year 2005 PDM Information,* at http://www.fema.gov/government/grant/pdm/fy2005.shtm.
[41] Interview with Michael Grimm, Mitigation Directorate, May 14, 2008.
[42] For details on listed programs, see CRS Report R40246, *Department of Homeland Security Assistance to States and Localities: A Summary and Issues for the 111th Congress*, by Shawn Reese.
[43] 42 U.S.C. 5133(c).
[44] 42 U.S.C. 5165(e).
[45] U.S. Department of Homeland Security-Federal Emergency Management Agency, *Fiscal Year 2006 and 2007, Pre-Disaster Mitigation Programs,* at http://www.fema.gov/government/grant/pdm/fy2006.shtm, *Fiscal Year 2007 Pre-Disaster Mitigation Program,* at http://www.fema.gov/government/grant/pdm/fy2007.shtm, and *Fiscal Year 2008 Pre-Disaster Mitigation Program,* at http://www.fema.gov/government/grant/pdm/fy2008.shtm . Note: Grants "selected for further review" refers to projects that have passed the first stage of review and await review for the National Environmental Policy Act (NEPA) and Environmental and Historic Preservation review. (Interview with Michael Grimm, FEMA Mitigation Directorate, May 13, 2008.)
[46] Ibid.
[47] FEMA updated the FY2007 amounts to $131 million expended for FY2007. This amount was $31 million over the appropriated amount for FY2007 and represents carry-over funding for projects that were selected in previous years but which had not received final approval. (Interview with Mike Grimm, FEMA Mitigation Directorate, May 22, 2008.)
[48] U.S. Department of Homeland Security - Federal Emergency Management Agency, *Fiscal Year 2006 Pre-Disaster Mitigation Program,* at http://www.fema.gov/government/grant/pdm/fy2006.shtm, and *Fiscal Year 2007 Pre-Disaster Mitigation Program,* at http://www.fema.gov/government/grant/pdm/fy2007.shtm.
[49] 42 U.S.C. 5133(g).

[50] Interview with Mike Grimm, Deputy Director, Risk Reduction Division, Mitigation Directorate, U.S. Department of Homeland Security, Federal Emergency Management Agency, May 20, 2008.
[51] U.S. Department of Homeland Security, Federal Emergency Management Agency, *Pre-Disaster Mitigation Program Guidance*, pp. 27-28 and pp. 39-40.
[52] U.S. Department of Homeland Security, Federal Emergency Management Agency, *Hazard Mitigation Grant Program Desk Reference*.
[53] Ibid.
[54] U.S. Department of Homeland Security, Federal Emergency Management Agency, "PDM Program Guidance, 4.3 Ineligible Program Activities and Costs," p. 40 at http://www.fema.gov/library/viewRecord.do?id=3029.
[55] P.L. 106-390, 114 Stat. 1557; P.L. 108-199, 118 Stat. 441; P.L. 108-447, 118 Stat. 3343; and P.L. 109-139, 119 Stat. 2649; P.L. 110-329, 122 Stat. 3690; and P.L. 111-83, 123 Stat. 2176.
[56] H.R. 1746.
[57] The trend continued in FY2008 with 117 of the 149 grant projects described as mitigation planning projects.
[58] Testimony of James Mullen, Mitigation Chair, National Emergency Management Association, in U.S. Congress, House Committee on Transportation and Infrastructure, Subcommittee on Economic Development, Public Buildings, and Emergency Management, *Saving Lives and Money Through the Pre-Disaster Mitigation Program*, hearing, 110th Cong. 2nd sess., April 30, 2008.
[59] Under H.R. 1746, passed by the House on April 22, 2009, the minimum would be increased to $575,000. H.R. 3377, passed by the House Transportation and Infrastructure Committee on November 5, 2009, also sets the minimum at that figure.
[60] U.S. Department of Homeland Security, Federal Emergency Management Agency, *HAZUS, FEMA's Methodology for Estimating Potential Losses from Disasters,* June 11, 2009 at http://www.fema.gov/plan/prevent/hazus/index.shtm.
[61] Department of Homeland Security, Appropriations Bill 2010, Report 111-XX, 111th Cong. 1st sess., p. 125.
[62] Michael J. Trebilcock and Ronald J. Daniels, "Rationales and Instruments for Government Intervention," in Ronald J. Daniels, Donald F. Kettl, and Howard Kunreuther, eds. *On Risk and Disaster: Lessons from Hurricane Katrina,* Philadelphia: University of Pennsylvania Press), p. 105.
[63] Testimony of Robert C. Bohlmann, U.S. Government Affairs Committee Chair, International Associations of Emergency Managers, in U.S. Congress, House Committee on Transportation and Infrastructure, Subcommittee on Economic Development, Public Buildings, and Emergency Management, *Saving Lives and Money Through the Pre-Disaster Mitigation Program,* hearing, 110th Cong. 2nd sess., April 30, 2008.
[64] U.S. Department of Homeland Security - Federal Emergency Management Agency, Pre-Disaster Mitigation Program Guidance, at http://www.fema.gov/library/viewRecord.do?id=3029.
[65] 42 U.S.C. 5132(g)(7).
[66] House Committee on Transportation and Infrastructure, Subcommittee on Economic Development, Public Buildings and Emergency Management, "Saving Lives and Money Through Pre-Disaster Mitigation," April 30, 2008.
[67] 42 U.S.C. 5133(g)(2).
[68] William L. Waugh, Jr., *Living With Disasters, Dealing With Disasters*, (New York: M.E. Sharpe, 2000), p. 155.
[69] 44 CFR Subpart A, 60.1(b).

[70] Rutherford H. Platt, *Disasters and Democracy: The Politics of Extreme Natural Events,* (Washington, DC: Island Press, 1999), p. 102.
[71] H.R. 2592, "Safe Building Code Incentive Act of 2009," 111th Cong. 1st sess. May 21, 2009.
[72] U.S. Department of Homeland Security - Federal Emergency Management Agency, Mitigation Directorate, Memo from FEMA Office of Legislative Affairs, July 16, 2008.
[73] The DRF is the no-year fund that funds disaster response and recovery programs. Congress provides funding both through annual appropriations and, most prominently, through supplemental appropriations to the DRF.
[74] U.S. Department of Homeland Security - Federal Emergency Management Agency, Mitigation Directorate, Grant Applications Guidance.
[75] U.S. Department of Homeland Security, Federal Emergency Management Agency, FY2009 Unified Hazard Mitigation Assistance (UHMA) Guidance, June 2008.
[76] U.S. Congress, House Committee on Appropriations, Department of Homeland Security Appropriations Bill 2009, 110th Cong., 2nd sess., H.Rept. 110-862, to accompany H.R. 6947, p. 109. After further negotiation, FEMA responded to this request with a briefing for Congressional staff. The briefing slides and information has helped to inform this chapter.

In: Federal Flood Policy
Editor: James E. Rysanek

ISBN: 978-1-61324-017-5
© 2011 Nova Science Publishers, Inc.

Chapter 3

FLOOD RISK MANAGEMENT AND LEVEES: A FEDERAL PRIMER[*]

Betsy A. Cody and Nicole T. Carter

SUMMARY

Midwestern flooding and Hurricane Katrina have raised concerns about reducing human and economic losses from flooding. In the United States, local governments are responsible for land use and zoning decisions that shape floodplain and coastal development; however, state and federal governments also influence community and individual decisions on managing flood risk. The federal government constructs some of the nation's flood control infrastructure, supports hazard mitigation, offers flood insurance, and provides emergency response and disaster aid for significant floods. In addition to constructing flood damage reduction infrastructure, state and local entities operate and maintain most of the flood control infrastructure and have initial flood-fighting responsibilities.

Prior to the Lower Mississippi River Flood of 1927, the federal role in flood control was limited. The Flood Control Act of 1936 (19 Stat. 1570) declared some flood control a "proper" federal activity. Today, the federal agencies most involved in flood control and flood fighting and emergency response are the U.S. Army Corps of Engineers (Corps) and the Federal Emergency Management Agency (FEMA).

---

[*] This is an edited, reformatted and augmented version of a Congressional Research Services publication, dated June 20, 2008.

The 110th Congress is faced with numerous flood control issues, including responding to disasters and adjusting federal flood policies. The recent midwestern floods and Hurricane Katrina have broadened interest in fundamental review of the current approach to managing floodwaters. Questions raised are: Do current policies, programs, and practices result in an acceptable level of aggregate national risk? Do they promote wise use and investments in the nation's floodplains and coasts? Do they encourage development that puts people in harm's way? Levees represent a particular challenge in that they may encourage development in flood-prone areas, but sometimes fail or are overtopped by significant storms. Hurricane Katrina brought national attention to the catastrophic consequences when structures fail or are breached. Similarly, two major midwestern floods in the span of 15 years (one in 1993 and one in 2008) have raised concerns about structures' ability to reduce or avoid flood damages and their effects on development patterns.

The 110th Congress addressed some flood issues in the first omnibus Water Resources Development Act (WRDA) enacted after Hurricane Katrina — WRDA 2007 (P.L. 110-114). For example, WRDA 2007 requires that national water resources planning avoid the unwise use of floodplains and flood-prone areas, and requires the President to report by 2010 on national vulnerability to flood damages, including the risk to human life. This chapter is to include assessments of current programs and recommendations for improvements. The law also creates a Committee on Levee Safety to make recommendations for a national levee safety program. How these changes are implemented over the next few years may affect the nature of federal investment in flood and storm damage infrastructure and mitigation measures.

This chapter provides a primer on responsibilities for flood management, describes the role of federal agencies, and discusses flood issues before the 110th Congress. The report also discusses the legislative response to Hurricane Katrina.

INTRODUCTION

Midwestern flooding in 2008 and Hurricane Katrina flooding in 2005 have enlivened interest in reducing the risk of flooding in communities across the nation. These large-scale events have demonstrated that not only is property damaged during floods, but also floods can represent significant risks to life and can cause economic disruption and other social hardships. The 110th Congress, like many earlier Congresses, is faced with numerous flood control issues, including responding to flood events and altering federal flood damage reduction, mitigation, and insurance policies. These issues have been brought to the fore as the Midwest experiences its second major flood in 15 years.[1]

In the United States, local governments are responsible for land use and zoning decisions that direct floodplain and coastal development; however, state and federal governments also influence community and individual decisions on managing flood risk. For example, the federal government constructs some of the nation's flood control infrastructure, supports hazard mitigation actions, offers flood insurance, and provides emergency response and disaster aid for significant floods. The federal agencies most involved in flood damage reduction and flood fighting and emergency response are the U.S. Army Corps of Engineers (Corps) and the Federal Emergency Management Agency (FEMA).

This chapter is divided into three sections. The first describes the current intergovernmental division of responsibilities for flood management and the federal role and interest in flood management. The second provides a framework for understanding flood risk management issues and the challenge of addressing the reliability and level of protection of the nation's levees. The third section describes actions that the 110th Congress has already taken and selected remaining issues that it, and many previous Congresses, have faced.

FLOOD MANAGEMENT RESPONSIBILITIES: A FEDERALIST DIVISION

Recent major flooding events have drawn attention to ongoing debates about how to improve management of flood risk and the roles and responsibilities of individuals, communities, and the various levels of government. As with many other policy areas, the federal system has resulted in public functions for flood damage reduction being shared by all levels of government. Local governments are responsible for land use and zoning decisions that direct floodplain and coastal development; however, numerous federal and state flood policies and programs influence local and individual decision-making. The federal government also funds some flood and storm damage reduction measures, manages a flood insurance and mitigation program, and provides disaster assistance.[2] It also generates essential data through mapping and other efforts.

Levees may be built by federal, state, or local entities (including private entities at the local level). Generally, levees are maintained by a local entity, with some exceptions. Local levee districts are generally the first entities responsible for monitoring levee conditions during flooding. The levee districts are also the first entity responsible for emergency response. If a flood or other emergency exhausts the levee district's flood fighting resources, the district typically contacts the state. The state will contribute its flood fighting resources to the local effort; as the state's resources are exhausted, it typically will contact the Corps for assistance under the Corps' emergency response authority.

Federal Role and Interest in Reducing Flood Damages

The federal role in flood control began in the late 19th century. Prompted by devastating floods in the Mississippi River basin, Congress created a commission to oversee the development of a levee system to control the river's flow. The Mississippi River Flood of 1927[3] and floods in the mid-1930s, ushered in a modern era of federal flood control investment. The Flood Control Act of 1936 (19 Stat. 1570) declared flood control a "proper" federal activity in the national interest.[4] Section 1 of the act established the following policy:

> It is hereby recognized that destructive floods upon the rivers of the United States, upsetting orderly processes and causing loss of life and property, including the erosion of lands and impairing and obstructing navigation, highways, railroads, and other channels of commerce between the States, constitute a menace to national welfare; that it is the sense of Congress that flood control on navigational waters or their tributaries is a proper activity of the Federal Government in cooperation with States, their political sub-divisions an localities thereof; that investigations and improvements of rivers and other waterways, including watersheds thereof, for flood-control purposes are in the interest of the general welfare; that the Federal Government should improve or participate in the improvement of navigable waters or their tributaries including watersheds thereof, for flood-control purposes if the benefits to whomsoever they may accrue are in excess of the estimated costs, and if the lives and social security of people are otherwise adversely affected.

As with many other policy areas, the federal system has resulted in public functions for flood damage reduction being shared by all levels of government. Since the mid-1980s, local project sponsors (often local governments or special levee and drainage districts) share construction cost of federal flood control projects and are fully responsible for operation and maintenance. Local entities (and sometimes state entities) may construct flood control infrastructure independently from the federal government, and are responsible for land use and zoning decisions guiding development in floodplains and coastal areas.

The impetus for federal and state attention to flooding comes from multiple sources. For instance, flooding often can occur regionally, and flood control works of one community can exacerbate or, alternatively, mitigate flood risk in other areas. Some federal and state actions attempt to alter individual and community behavior to account for flooding risks and losses. Most individuals discount the probability of loss from infrequent events, even if those events may cause significant losses and disruption. In general, many local decision makers do not view environmental hazards, such as flooding, as serious problems, in comparison to the many other problems that local governments are expected to address.[5]

Principal Federal Agencies

As previously noted, the Corps and FEMA are the principal federal agencies involved in flood damage reduction and flood fighting and emergency response. Other federal agencies also are involved with flood damage reduction projects, such as the U.S. Department of Agriculture's Natural Resources Conservation Service, the Department of the Interior's Bureau of Reclamation, and the Tennessee Valley Authority.

At the direction of Congress, the Corps is authorized to participate in the cost-shared planning and construction of flood damage reduction projects, such as building levees and floodwalls to reduce damages from coastal and riverine flood hazards. The Corps is responsible for much of the federal construction investment in flood control and storm protection infrastructure. It has constructed nearly 9,000 miles of the nation's roughly 15,000 miles of levees. Corps involvement in flood control construction is predicated on the project being in the national interest, which is determined by the likelihood of widespread and general benefits, a shortfall in the local ability to solve the water resources problem, the national savings achieved, and precedent and law.[6]

Generally, after construction by the federal government, this infrastructure is turned over to a local entity for operation, maintenance, repair, and rehabilitation. The Corps, however, has retained responsibility for roughly 900 miles of levees, primarily along the Mississippi River and for multi-purpose dams. FEMA has various programs, such as its Hazard Mitigation Grant Program and its Flood Mitigation Assistance Program, that promote flood mitigation actions, such as assisting in removing vulnerable structures from floodplains and other activities that reduce the impact of a flood disaster.

The Corps performs most of the federal inspections of levees. Levee inspections are conducted for participation in two federal programs. The first is the Corps' Rehabilitation and Inspection Program. This program provides federal assistance for repairing levees damaged during floods. The Corps is to conduct annual (or semiannual) inspections of levees for initial inclusion in the program and for continued eligibility for assistance. The Corps also often performs the inspections to certify a levee's reliability for a 100-year flood under FEMA's National Flood Insurance Program (NFIP).[7]

Congress gave the Corps emergency response authority that allows the agency to fight floods and other natural disasters. P.L. 84-99 (33 U.S.C. §701n) provides the Corps authority for emergency response and disaster assistance. It authorizes disaster preparedness, advance measures, emergency operations (disaster response and post-flood response), rehabilitation of flood control works threatened or destroyed by floods, protection or repair of federally authorized shore protection works threatened or destroyed by coastal storms, emergency dredging, and flood-related rescue operations. These activities are limited to actions to save lives and protect improved property (public facilities/services and residential or commercial developments).[8] FEMA can also direct the Corps and other agencies to undertake activities in response to flooding and other national emergencies, as part of FEMA's implementation of the National Response Framework.[9]

A FLOOD RISK FRAMEWORK

Hurricane Katrina and recent midwestern flooding demonstrate that not only property damage but also significant risks to life, economic disruption, and other social hardships occur during floods. Flood risk is a composite of three factors:

- *vulnerability*, which allows a threat to cause consequences (e.g., level of protection provided by levees and dams, their reliability, and location within a floodplain);[10]
- *threat* of an event (e.g., probability of a Category 5 hurricane storm surge or a 200-year flood affecting a particular location); and
- *consequence* of an event (e.g., property damage, loss of life, economic loss, environmental damage, reduced health and safety, and social disruption).

Reducing Vulnerability and the 100-Year Flood

In the United States, the 1% annual chance flood, more commonly known as the 100-year flood, is a standard often used as a basis for identifying, mapping, and managing flood hazards. For example, the NFIP and most state and local governments use location in the 100-year floodplain or similar coastal zone inundation areas as triggers for various requirements. The 100-year flood standard was established at the recommendation of a group of experts in the late 1960s. "It was selected because it was already being used by some agencies, and it was thought that a flood of that magnitude and frequency represented a reasonable probability of occurrence and loss worth protecting against and an intermediate level that would alert planners and property owners to the effects of even greater floods."[11] The adoption of the 100-year flood standard in many respects guides perceptions of what is an *acceptable level of vulnerability*. The 100-year flood standard is a vulnerability standard, and not a risk standard. Thus, the question of whether the 100-year flood standard combined with current threat and consequence information results in an *acceptable level of risk* remains largely unaddressed; this question is especially relevant for low probability, high consequence events such as a Category 4 hurricane hitting a major urban center.

The attempt to provide at least 100-year flood protection often drives local floodplain management and infrastructure investments, resulting in a measure of equity within and across communities. That equity in vulnerability, however, results in uneven levels of risk because flooding of different communities has different consequences, such as differences in the potential loss of life, social disruption, structures damaged, and economic impact because of variations in land use and development patterns.

The National Flood Insurance Program does not differentiate between 100-year flood protection provided by a flood control structure and flood protection resulting from natural topography and hydrology. As a result, development behind levees and downstream of dams providing 100-year flood protection is not designated as located in a "special flood hazard area," thus freeing occupants from flood insurance requirements. While the NFIP largely presumes that levees, dams, and other flood control structures will not fail, their presence does not entirely eliminate an area's vulnerability to flooding.

The *residual flood risk* behind levees or downstream of dams remains largely unaccounted for in the NFIP and often is not incorporated into individual, local, and state decision-making. Residual risk is the portion of risk that remains after flood control structures have been built and other damage-reducing measures have been taken. Risk remains because of the likelihood of the measures' design being surpassed by floods' intensity and of structural failure of the measures. Often when the designs of flood control structures are surpassed or when structures fail for other reasons, the resulting flood is catastrophic, as shown by the floodwall breaches in New Orleans (LA) with Hurricane Katrina in 2005. The damaging consequences of floods increase as development occurs behind levees and below dams; ironically, this development may occur because of the flood protection provided. The nation's risk in terms of lives lost, economic disruption, and property damage is increased by overconfidence in the level and reliability of structural flood protection.

Next Step: A Risk Management Approach?

Investments in flood control measures, such as dams and levees, and emergency response activities have resulted in a decreasing trend (excluding the deaths associated with Hurricane Katrina and most recent midwestern floods) in lives lost to flooding since the 1920s; during the same period, property damage due to flooding has been increasing. Through the NFIP, the federal government attempts to promote flood-hazard awareness and damage-reducing practices, as well as to assist individuals in managing flood losses. While this produces clear benefits for moderate floods, some stakeholders are concerned that structural flood control measures and the NFIP together may contribute to a false sense of security for individuals and communities. This sense of security may foster decisions to locate in potentially hazardous areas, thus increasing the national vulnerability to flood losses.

The 2008 midwestern floods and Hurricane Katrina have contributed to interest in fundamental reexaminations of the approach to managing floodwaters. Some of the questions raised are: Do current policies, programs, practices, and investments result in an acceptable level of aggregate risk for the nation? Do they promote wise use and investments of the nation's floodplains and coasts?

Risk management is being increasingly viewed as a method for setting priorities for managing some hazards in the United States. Because floodplain and coastal development are largely managed by local governments, some aspects of national flood risk management likely would be unwelcome and infeasible, and could be perceived as resulting in an inequitable distribution of flood protection. For example, if floods in large urban concentrations are perceived as representing a greater risk for the nation, federal resources may be directed away from protecting smaller communities and less-populated states. Two of the concerns raised in discussions of greater emphasis on risk analysis in the development and design of specific projects are that risk analysis may result in lower levels of protection being implemented in some areas, and that information and knowledge are insufficient to perform an adequate analysis. However, an argument can be made that the federal government has an interest in reducing risks resulting in national consequences, and in prioritizing federal involvement and appropriations accordingly.

Factors complicating the determination of the nation's flood risk include changing conditions and incomplete information. For example, many flood control projects were built decades ago using the available data, technologies, and scientific knowledge of the period that may have underestimated flood hazards for particular areas. Similarly, there are issues with changes in risk over time due to processes such as land loss, subsidence, sea-level rise, reduced natural buffers, urban development, and infrastructure aging. For existing dams, there is some information on consequences of failure as measured by loss of life, economic loss, environmental loss, and disruption of lifeline infrastructure (such as bridges and power grids); however, the database with this information only tracks the amount and type of losses, not the likelihood of failure.[12]

A risk-reduction approach for organizing federal flood-related investments likely would incorporate many structural and nonstructural flood management measures already being considered and implemented, but change their priority and mix. Options considered in a risk-centered approach may include shifting federal policy toward wise use of flood-prone areas (e.g., rules or incentives to limit some types of development in floodplains), incorporating residual risk and differences in riverine and coastal flood risk into federal programs (e.g., residual risk premiums as part of the National Flood Insurance Program), creating a national inventory and inspection program for levees, promoting greater flood mitigation and damage mitigation investments, re-evaluating operations of flood control reservoirs for climate variability and uncertainty, and investing in technology and science for improved understanding of flooding threats.

The Levee Challenge

Hurricane Katrina brought national attention to the issue of levee and flood wall reliability and different levels of protection provided by flood damage reduction structures, particularly those protecting concentrated urban and population centers. A 1982 National Research Council report stated that levee overtopping or failure was estimated to be involved in approximately one-third of all flood disasters, and that the nation's dam inspection program suggests that a large percentage of locally built levees are likely poorly designed and maintained.[13] How to address levee reliability and various levels of protection remains at issue.

Many levees protecting today's communities and agricultural investments originally were planned and constructed beginning nearly a century ago (or more than a century ago) by local interests attempting to reclaim land to make it productive for agriculture and other uses. Rather than each landowner building separate levees, landowners often consolidated their resources by forming a levee district. As a consequence of this history, many of today's physical constructions and configurations, as well as institutional arrangements, for flood protection have roots distinct from their current use as flood protection for development. Most levees currently are operated by a levee district or some other special or general local government. For the most part, municipalities serving concentrated urban populations have assumed flood control responsibilities, while special levee districts remain abundant in rural and agricultural areas. Note, however, that there are exceptions to this generality.

An issue that may limit government entities' interest in levee construction, maintenance, and possibly inspection responsibilities is liability for flood damages. A principal source of concern may stem from the uncertainty related to the implications of *Paterno v. State of California*, which held the State of California liable for a levee it did not build, but operated as part of a state-sponsored levee system.[14] The issue of federal liability for damages is discussed in CRS Report RL34131, *Federal Liability for Hurricane Katrina-Related Flood Damage*, by Cynthia Brougher and Kristina Alexander.

FLOOD MANAGEMENT ISSUES IN THE 110TH CONGRESS

A Legislative Response to Katrina's Lessons: Factoring in Safety

In the first omnibus Water Resources Development Act (WRDA, which is the legislative authorization vehicle for the Corps) enacted after Hurricane Katrina — WRDA 2007 (P.L. 110-114) — Congress addressed a number of policy changes and authorized numerous flood and storm damage reduction projects and project modifications. WRDA 2007 included the following provisions specifically related to flood-related policies:

- *Water Resources Principles and Guidelines* (§2031) — This provision states a national water resources planning policy that includes avoiding unwise use of floodplains and flood-prone areas, and requires the Corps to update by 2010 the guidelines it uses for planning and implementing Corps water resources projects.
- *Water Resources Priorities Report* (§2032) —Ths provision requires the President submit to Congress a report by 2010 on the vulnerability of the nation to flood damages, including the risk to human life, which is to include assessments of current programs and recommendations for improvements.
- *Planning* (§2033) — This provision makes changes to Corps planning activities, including requirements that the economic analysis of flood damage reduction projects consider the risk that remains behind levees and floodwalls, upstream and downstream impacts, and equitable analysis of structural and nonstructural alternatives.
- *Safety Assurance Review* (§2034) — This provision requires that the design and construction of Corps flood and storm damage reduction projects be independently reviewed by experts to assure public health, safety, and welfare.
- *National Levee Safety Program* (Title IV) — This title creates a Committee on Levee Safety to make recommendations to Congress by mid-2008 for a national levee safety program; however, the committee has not yet been funded. The title also requires the Corps to establish and maintain a database with an inventory of the nation's levees by 2009 and to inspect federally constructed and other levees.

How these changes are implemented over the next few years may affect the nature of the federal investment in flood and storm damage infrastructure and mitigation measures.

Selected Remaining Issues

The 2005 hurricane season and the 2008 midwestern floods have focused the nation's attention once again on issues that flood experts have debated for decades. The devastation of these events renewed public concerns about reliability of the nation's aging flood control levees and dams. The debate over what is an acceptable level of risk — especially for low-probability, high-consequence events — and who should bear the costs to reduce the flood risk (particularly in the case of levees) is taking place not only in the affected states, but nationally. The concerns being raised range widely, including interest in providing more protection for concentrated urban populations, risk to the nation's public and private economic infrastructure, support for reducing vulnerability by investing in natural buffers, equity in protection for low-income and minority populations, consistency in and the form of flood insurance and disaster aid, and the level of federal, state, and local investment in structural and nonstructural flood damage reduction measures.

Response to the 2005 hurricane season and previous midwestern floods included discussions of expanding mitigation activities (such as floodproofing structures and buyouts of structures on the most flood-prone lands), investing in efforts to restore natural flood and storm surge attenuation, and assuring vigilant maintenance of existing flood control structures, as well as interest in new and augmented structural flood protection measures. Although major flood events, generally spur these discussions, the policy changes implemented often are incremental.[15] The 110th Congress, like previous Congresses, faces a challenge in reaching consensus on whether and how to proceed on anything other than incremental change because of the wealth of constituencies and communities affected by federal flood policy. Another practical challenge is the division of congressional committee jurisdictions over the federal agencies and programs involved in flood mitigation, protection, and response.[16]

There are many questions that remain about how events unfolded in the aftermath of Hurricane Katrina, and much information that is still needed to understand how to apply and communicate nationally the lessons in the Gulf and midwestern states learned about flood risk and disaster preparedness and response. Although there is no way to protect against all flood risk, many contend that more information is needed to evaluate flood risk, to understand the reliability and residual risk of structural flood protection, and to incorporate the full range of flood consequences into local, state, and federal decision-making and programs.

End Notes

[1] Major flooding in the Midwest is reported to be in the range of a 400-year to 500-year flood; however, most levee protection is built to withstand a 100-year flood. These flood-year designations, however, do not indicate how often an area may flood. Rather, they are based on the chance that an area may flood in any given year. For example, the term *100-year flood* is the flood elevation that has a 1% chance of being equaled or exceeded *annually*. It is *not* the flood that will occur once every 100 years; 100-year floods can occur more than once in a relatively

short period of time. Likewise, a 500-year flood is five times less likely to occur in any given year then a 100-year flood (0.2% chance of flooding).

[2] For information on the evolution of federal disaster aid, see U.S. Senate Task Force on Funding Disaster Relief, *Federal Disaster Assistance*, S.Doc. 104-4 (1995). For information on federal programs providing disaster assistance, see the CRS Disaster Assistance and Recovery Web page at [http://apps.crs.gov/cli/cli.aspx?PRDS_CLI_ITEM_ID=2432].

[3] For more information on the response to the Mississippi River Flood of 1927, see CRS Report RL33126, *Disaster Recovery and Appointment of Recovery Czar: The Executive Branch's Response to the Flood of 1927*, by Kevin R. Kosar.

[4] The Beach Nourishment Act of 1956 (P.L. 84-826) expanded the federal role in constructing projects for hurricane, storm and shoreline protection, such as seawalls and the periodic placement of sand on beaches to control erosion. The Flood Control Act of 1950 (64 Stat. 170) began the Corps' emergency operations by authorizing flood preparedness and emergency operations.

[5] R. Burby, "Hurricane Katrina and the Paradoxes of Government Disaster Policy," prepared for *Annals of the American Academy of Political and Social Science* (March 2006).

[6] This is described in the Corps' *Digest of Water Resources Policies and Authorities* Engineering Pamphlet EP 1165-21-1 (1999).

[7] As discussed earlier, the term *100-year flood* is the flood elevation that has a 1% chance of being equaled or exceeded *annually*. It is *not* the flood that will occur once every 100 years; 100-year floods can occur more than once in a relatively short period of time. The 1994 "Galloway Report" (see note 15) uses an analogy of a bag of 100 marbles where 99 are clear and 1 is black. Every time you pull out a black marble would be equivalent to a 100-year flood, but the black marble is replaced and the bag is shaken up before you draw again. So, it is possible, but not likely, you might draw the black marble two or three times in a row or with greater frequency than only one time in 100 draws.

[8] Although the Corps' account paying for these activities may receive some appropriations in the annual Energy and Water Development Appropriations acts, this initial appropriation is often supplemented with emergency appropriations specific to the emergency being addressed.

[9] For more information, see CRS Report RL33053, *Federal Stafford Act Disaster Assistance: Presidential Declarations, Eligible Activities, and Funding*, by Keith Bea.

[10] For more information on this three-part hazard risk framework, see CRS Report RL32561, *Risk Management and Critical Infrastructure Protection: Assessing, Integrating, and Managing Threats, Vulnerabilities, and Consequence*, by John Moteff.

[11] Association of State Flood Plain Managers, *Reducing Flood Losses: Is the 1% Chance (100-year) Flood Standard Sufficient?* (Washington, DC: 2004).

[12] For information on dam safety, see CRS Report RL33108, *Aging Infrastructure: Dam Safety*, by Nic Lane.

[13] National Research Council, A Levee Policy for the National Flood Insurance Program, (U.S. Dept. of Commerce: Oct. 1982).

[14] *See* Paterno v. State of California, 2003 Cal. App. LEXIS 1771 (2003) *pet. for rev. denied*, 2004 Cal. LEXIS 2253 (Mar. 17, 2004); see also Arreola v. County of Monterey 2002 Cal. App. LEXIS 4319 (2002) *pet. for rev. denied*, 2002 Cal. LEXIS 6194 (Sept. 18, 2002).

[15] After the Midwest Flood of 1993, the Interagency Floodplain Management Review Committee was directed to evaluate the performance of floodplain management and make recommendations in current policies and programs of the federal government. The resulting 1994 report, titled *Sharing the Challenge: Floodplain Management in the 21st Century*, often called the "Galloway

Report," for the Committee's chair, includes the Committee's recommendations; the report is available at [http://eros

[16] Several different congressional committees could potentially claim jurisdiction over elements of comprehensive change in federal flood policy. For a discussion of jurisdictional issues in the House, see CRS Report 98-175, *House Committee Jurisdiction and Referral: Rules and Practice,* by Judy Schneider; for Senate jurisdiction, see CRS Report 98-242, *Committee Jurisdiction and Referral in the Senate,* by Judy Schneider.

In: Federal Flood Policy
Editor: James E. Rysanek

ISBN: 978-1-61324-017-5
© 2011 Nova Science Publishers, Inc.

Chapter 4

MIDWEST FLOODING DISASTER: RETHINKING FEDERAL FLOOD INSURANCE?[*]

Rawle O. King

SUMMARY

Historically, floods have caused more economic loss to the nation than any other form of natural disaster. In 1968, Congress created the National Flood Insurance Program (NFIP) in response to rising flood losses and escalating costs resulting from ad-hoc appropriations for disaster relief. Federal flood insurance was designed to provide an alternative to federal disaster relief outlays by reducing the rising federal costs through premium collection and mitigation activities. The purchase of flood insurance was considered to be an economically efficient way to indemnify property owners for flood losses and internalize the risk of locating investments in the floodplains.

Despite massive rainfall-river flooding in several Midwestern states along the upper Mississippi River and its tributaries in June 2008, damages for the most part are not expected to produce significant insured flood losses under the NFIP. This significant but not unprecedented flood event instead will likely cost several billions in uninsured damages that will probably remain uncompensated or be paid through federal emergency supplemental appropriations for disaster relief.

---

[*] This is an edited, reformatted and augmented version of a Congressional Research Services publication, dated August 25, 2008.

A key lesson learned from the 1993 and 2008 Midwest floods is that many people believe that the government will provide them with economic assistance despite their lack of insurance. What then is the appropriate role of the federal government in dealing with ambiguous risks, where the insurance industry is reluctant to offer coverage and homeowners and businesses demonstrated a reluctance to purchase coverage, even when it is mandatory? This question is important for the long-term solvency of the NFIP and overall future costs to federal taxpayers.

This chapter examines the impact of the 2008 Midwest floods on the National Flood Insurance Program (NFIP) in the context of congressional efforts to reauthorize and modify the program before its authorization expires on September 30, 2008. The report begins with an assessment of the risk of flooding in the United States and why Congress might move to rethink the current multifaceted approach to federal flood insurance. Members might, for example, opt to assess possible insurance requirements for individuals living behind levees, eliminate premium subsidization of certain "grandfathered" properties, expand the NFIP to offer coverage against both flood and wind damages, and consider undertaking a nationwide flood insurance study (FIS) and remapping of the nation's floodplains, including areas behind levees and other flood control structures. The report concludes with lessons learned from the 1993 and 2008 Midwest floods, and an analysis of the NFIP's current financial conditions and major policy issues, as well as a summary of legislative proposals — H.R. 3121 and S. 2284 — pending before the 110th Congress.

INTRODUCTION

This chapter examines the impact of the 2008 Midwest floods on the National Flood Insurance Program (NFIP) in the context of congressional efforts to reauthorize and modify the program before its authorization expires on September 30, 2008. The report begins with an assessment of the risk of flooding in the United States and why Congress might move to rethink the current approach to federal flood insurance. Members might, for example, opt to assess possible insurance requirements for individuals living behind levees, eliminate premium subsidization of certain "grandfathered"properties, expand the NFIP to offer coverage against both flood and wind damages, and consider undertaking a nationwide flood insurance study (FIS) and remapping the nation's floodplains, including areas behind levees and other flood control structures. The report concludes with lessons learned from the 1993 and 2008 Midwest floods and 2005 hurricanes, and an analysis of the NFIP's current financial conditions and major policy issues, as well as a summary of legislative proposals — H.R. 3121 and S. 2284 — pending before the 110th Congress.

Despite massive rainfall-river flooding in several Midwestern states along the upper Mississippi River and its tributaries in June 2008, damages for the most part are not expected to produce significant insured flood losses under the National Flood Insurance Program (NFIP). This extensive but not unprecedented flood event instead will likely cost several billions in uninsured damages that will probably remain uncompensated or be paid through federal emergency supplemental appropriations for disaster relief.

Relatively few NFIP claims (6,338) had been filed as of June 30, 2008. Insurance policies sold by private insurers generally do not insure for the flood peril. Without federal flood insurance or private insurance, flood victims typically finance flood damage repair costs on their own (self-insure), claim a tax credit for property loss on their individual returns and, in the event of a presidentially declared major disaster, pay a portion of the uninsured losses with federal disaster relief assistance.

Federal assistance is usually provided to eligible individuals and businesses under section 408 of the Robert T. Stafford Disaster Relief and Emergency Assistance Act, as amended by the Disaster Mitigation Act of 2000.[1] Flood-prone residents might also be eligible for voluntary buyouts under the Hazard Mitigation Grant Program (HMGP) that funds property acquisitions to mitigate future flood disaster losses.[2] After the 1993 Midwest floods, approximately 12,000 properties in nine Midwestern states were bought out by the government and about 500 other structures were relocated or elevated. Buyouts are again being considered in five state affected by the 2008 floods: Missouri, Iowa, Wisconsin, Indiana and Illinois.[3]

Occurring less than three years after the widespread devastation — floods, storm surge and breached levees — caused by Hurricanes Katrina and Rita, the 2008 Midwest flood has once again brought to the forefront of public awareness weaknesses in the nation's flood management system. The 2008 floods have also focused public attention on the lack of understanding of the national flood risk, uncoordinated federal flood risk programs, diminished capabilities in flood risk management, outdated floodplain information (flood hazard maps), and the flood damage destruction that can occur when levees are breached or overtopped.[4]

The next two sections of the report provide an assessment of the U.S. risk of flooding and why Congress might decide to evaluate the current approach to federal flood insurance. This is followed by an analysis of lessons learned from the 1993 Midwest floods, the financial status of the program after the first catastrophic floods in the program's history, and the policy issues that emerged from apparent weaknesses highlighted by the 2005 hurricanes.

THE U.S. RISK OF FLOODING

Historically, flooding has caused more economic loss to the nation than any other natural hazard. Flooding in the United States has been a recurring event, and the severity of flooding varies from year to year and from location to location. Almost 90% of all declared disasters include a flooding component. Flooding is not confined to just a few geographic areas.[5] The Midwestern floods in June 2008 that occurred along the upper Mississippi River and its tributaries demonstrate that flood affects can be local, impairing a neighborhood or community, or very large, affecting entire river basins or multiple states. Despite the billions of dollars that have been spent for structural flood control and FEMA's multifaceted approach to mitigating property losses, flood-related damages continue to rise.[6]

The magnitude of flood events has traditionally been measured by recurrence intervals, or the likelihood that a flood of a particular size will recur during any 10-, 50-, 100-, or 500-year period. These events have a 10-, 2-, 1-, and 0.2-percent chance, respectively, of being equaled or exceeded during any year. (Rare floods sometimes occur at short intervals or even within the same year.)

The sources of the nation's rising flood risks are many. Increased urbanization and coastal development have reportedly led to both heightened exposure of people and property along rivers and greater chances of flood losses. The Government Accountability Office has recently reported that weather-related events have cost the NFIP billions in damages, and suggested that climate change may increase losses due to increased frequency or severity of weather-related events.[7] Climate scientists with the U.S. Climate Change Science Program of the National Oceanic and Atmospheric Administration (NOAA) predict that, due to global warming, severe precipitation events that once occurred every 20 years in many parts of the country could happen once every 4 to 6 years by the end of the 21st century.[8]

Congressional interest in flooding and flood control policy originated in the late 19th century, following massive flooding along the Mississippi River basin during the 1850s through 1870s when policymakers began to consider strategies to mitigate the escalating costs of repairing damage to buildings and their contents caused by floods. The federal policy response to widespread flood damages in communities and rising taxpayer-funded disaster relief cost was initially a so-called "levee-only" policy approach — i.e., relying on levees to protect population and property in flood-prone areas. In 1879, Congress created the Mississippi River Commission (1879-1928) to oversee the development of a levee system that would confine the river's natural flow.[9] Since the enactment of the Flood Control Act of 1917, the U.S. Army Corps of Engineers (USACE) has played a significant role in flood damage reduction.[10]

Over the 40-year period between the historic Mississippi Floods of 1927 and the early 1960s, it became generally apparent that the "levee-only" flood control policy approach was not achieving the intended objectives.[11] This strategy of modifying nature's flood hazard areas would prove costly. Largely because of Hurricane Betsy (1965) and other hurricanes in 1963 and 1964, as well as heavy flooding on the upper Mississippi River basin in 1965, Congress undertook a study of the feasibility of alternative methods of providing assistance to those suffering property losses in floods and other natural disasters.[12] The recommendations of this study and private insurers' unwillingness or inability to underwrite flood insurance led to the enactment of the National Flood Insurance Act of 1968[13] and creation of the NFIP.

Flood damage is excluded under homeowner's policies because insurers consider flood risk to be an uninsurable peril. Insurers reportedly cannot accurately estimate losses and most lack the ability to pool and spread flood risks over a large and diverse group of (uncorrelated) insureds in order to minimize the possibility of multiple claims for the same event. Federally backed flood insurance fills this void and is available for residential and commercial properties in participating NFIP communities.

IS IT APPROPRIATE TO RETHINK FEDERAL FLOOD INSURANCE?

Midwestern flooding in 2008 caused dozens of levees to be breached, destroying thousands of homes and businesses, and inundated many thousands of acres of cropland. The flooding has once again focused public attention on the economics of government risk-bearing (federal flood insurance) when private insurers do not offer affordable coverage, on the exposure of federal taxpayer to losses when program revenues do not cover costs, and on the efficacy of the nation's floodplain management strategy in reducing federal disaster relief expenditures.

The Midwest floods also raised several broad issues and concerns that could lead policymakers to rethink federal flood insurance in the context of reauthorizing the NFIP. Broadly speaking, these issues and concerns include:

- What responsibilities should property owners bear to understand and prepare for flood hazards, especially given the confluence of greater property exposure and a projected likelihood of more frequent severe storms? How do we provide assistance to victims of floods? Is the NFIP the appropriate structure for insuring flood losses? Should Congress consider a comprehensive natural disaster program?
- Effectiveness of structural (levees and dams) and non-structural (land-use ordinances and building codes) floodplain management systems that annually prevent billions of dollars in flood-related property damages, but also arguably can encourage individuals and businesses to build in flood prone areas.
- The persistently low take-up rate of flood insurance in high-risk areas despite federal mandatory purchase requirements (lender compliance) for properties in a federally mapped flood zone.
- Whether households are relying on federal disaster relief for compensation, rather than federal flood insurance that Congress established 40 years ago to reduce such relief?

Background on the NFIP

In 1968, Congress created the NFIP in response to rising flood losses and escalating costs to the general taxpayers for disaster relief. Federal flood insurance was designed to provide an alternative to federal disaster relief outlays.[14] The purchase of flood insurance was considered to be an economically efficient way to indemnify property owners for flood losses and have them internalize some of the risk of locating investments in the floodplains.[15]

The NFIP is administered by the Federal Emergency Management Agency (FEMA) and provides subsidized and actuarially priced flood insurance policies for individuals, businesses, and renters located within and outside designated floodplains. The NFIP is a quid pro quo program in that FEMA agrees to make federally backed flood insurance available only in communities that agree to adopt and enforce floodplain management ordinances designed to reduce the future vulnerability of the built environment. The federal government retains responsibility for all underwriting losses, but it also has advantages over private insurers — namely, its greater ability to avoid adverse selection and moral hazard through mandatory purchase requirements (compulsory membership) and its access to greater information (risk assessment and flood hazard mapping).

Recognizing the low market penetration of flood insurance in the early 1970s, Congress enacted the Flood Disaster Protection Act of 1973[16] to establish a mandatory flood insurance purchase requirement for structures located in identified special flood hazard areas (SFHA).[17] The idea was to shift more of the cost of floods to those who build in flood-prone areas. After the 1993 Midwest floods, it became apparent that homeowners were still not adequately complying with the mandatory purchase requirement. The National Flood Insurance Reform Act of 1994 was enacted to strengthen the purchase requirement.[18] In 2004, Congress enacted the Flood Insurance Reform Act of 2004 to address, among other things, the repetitive loss property (RLP) problem.[19]

Challenges Facing the NFIP

In the wake of the 2005 hurricanes and the 2008 Midwest floods, critics of the status quo have pointed to several recurring problems that they believe affect the NFIP. They include:[20]

- The lack of accounting by the NFIP for residual risk behind levees has contributed property owners dismissing their exposure to risk of levee failure and overtopping. The 1993, 2005, and 2008 events illustrate how ignoring residual risk could contribute to greater demand for disaster relief.
- Inaccurate flood maps that need updating to reflect not only recent development in flood-prone areas but residual flood risk behind levees, dams and other structural flood control systems.
- Providing disaster relief to uninsured individuals tends to discourage steps to reduce loss exposure and results in higher societal and federal costs.
- The low flood insurance market penetration in river flood-prone areas that results in billions of dollars in uninsured losses.
- Substantial cross-subsidies among classes of policyholders with the use of "historical average loss year" premium setting approach.
- The performance of FEMA floodplain management standards in achieving flood damage reduction.

- The desirability of establishing a catastrophe reserve fund to pay claims during the rare catastrophic loss year. A reserve fund would mean higher premium rates.

MIDWEST FLOODS OF 2008

Early rough estimates of flood damages from the 2008 Midwest floods indicated that the cost to the NFIP would likely be small because of the relatively low number of policies and claims filed in Midwestern states. Most experts appear to agree the program will be able to cover 2008 flood claims without having to borrow from the Department of the Treasury. The NFIP might still have to borrow to pay scheduled interest payments on the debt, however.

Table 1 shows the number of federal flood insurance policies and the number of total claims submitted in 12 Midwestern states, as a result of the March and June 2008 floods. The table also shows total claims filed and total payments over the 20 year period from 1988 to 2007. NFIP had reportedly received only about 6,338 insurance claims, as of June 30, 2008. Iowa has experienced the majority of damages based on the number of counties affected (83 of 99 counties), policies in force, coverage in force, severity of damage to residential and commercial structures, characteristics of the flooding, and the *estimated* or average amount of the claim payment.

Table 2 provides a list of the top fifteen significant flood events in the United States in terms of NFIP payouts. The 2008 Midwest flood does not rank among these. Although the 1993 Midwest flood was the most devastating in the region, with total economic damages approximately $20 billion, it ranks only 12th in terms of the NFIP, with only $273 million in NFIP claims. In contrast, the devastating flooding caused by Hurricanes Katrina, Rita and Wilma resulted in more than $200 billion in economic losses of which $21.9 billion were covered under the NFIP.

The key lesson from the 2008 Midwest floods is not the magnitude of payouts under the NFIP but, rather, the eventual cost of federal disaster relief for individual, business and communities. Congress might therefore opt to reassess federal flood insurance that was intended to work in tandem with risk identification /mapping and floodplain management regulation to reduce flood losses.

Table 1. Federal Flood Insurance Policies Issued and Claims Paid to Midwestern States: 1988-2007 and June 2008

State	# of Policies (As of 4/30/08)	Claims Reported[a] (As of 6/30/08)	Total Claims	Total Payments	Rank
Illinois	48,404	691	11,422	$151,794,472	3
Indiana	29,091	1,418	5,577	72,696,096	6
Iowa	10,930	2,584	4,382	62,762,314	7
Kansas	11,995	0	2,916	60,050,079	8
Michigan	25,838	23	1,496	17,636,680	10
Minnesota	8,624	0	5,378	94,988,744	5
Missouri	24,223	364	15,560	327,648,567	1
Nebraska	11,821	0	1,102	12,239,363	12
North Dakota	4,431	0	5,482	124,358,171	4
Ohio	40,745	3	10,330	192,508,059	2
South Dakota	3,130	0	1,123	15,160,056	11
Wisconsin	13,958	1,281	2,615	29,507,952	9
Total	233,190	6,338	67,383	$1,161,350.553	NA

The first two data columns are grouped under "NFIP Policies Issued and Claims Reported"; the last three under "Total Claims Payments: 1988-2007 (Constant Dollars, As of 1/31/08)".

Source: U.S. Department of Homeland Security, Federal Emergency Management Agency.
a Includes claims filed March-June 2008. In addition to these 12 Midwestern states, three states experienced flood losses in March that resulted in NFIP claims: Arkansas (439), Kentucky (195), and Texas (106).

Table 3 illustrates that none of the 12 Midwestern states made the list of the top 10 states ranked by flood insurance claims payments made under the NFIP over the 20 year period from 1988 to 2007.

Table 2. Top Fifteen Significant Flood Events Covered in the National Flood Insurance Program (1988- April 30, 2008)

Rank	Event	Date	Number of Paid Losses	Amount Paid ($ Constant)	Average Paid Loss
1	Hurricane Katrina	Aug. 2005	164,917	$15,920,395,412	$95,077
2	Hurricane Ivan	Sep. 2004	27,304	1,566,138,612	55,518
3	Tropical Storm Allison	Jun. 2001	30,627	1,103,696,091	35,944
4	Louisiana Flood	May 1995	31,343	585,067,886	18,667
5	Hurricane Isabel	Sep. 2003	19,685	490,643,154	24,076
6	Hurricane Floyd	Sep. 1999	20,438	462,270,253	22,614
7	Hurricane Rita	Sep. 2005	9,328	458,251,687	47,428
8	Hurricane Opal	Oct. 1995	10,343	405,528,543	39,208
9	Hurricane Hugo	Sep. 1989	12,843	376,494,566	29,315
10	Hurricane Wilma	Oct. 2005	9,530	361,259,895	37,340
11	Nor'easter	Dec. 1992	25,141	346,151,231	13,768
12	Midwest Flood	Jun. 1993	10,472	272,827,070	26,053
13	PA, NJ, NY Floods	Jun. 2006	6,386	224,237,061	35,114
14	Nor'Easter	Apr. 2007	8,603	222,735,529	25,890
15	March Storms	Mar. 1993	9,841	212,616,751	21,605

Source: U.S. Department of Homeland Security, Federal Emergency Management Agency.

Tables 2 and 3 indicate that hurricanes — not river flooding — have been the major natural hazard contributing to NFIP claims payments and the most significant flood events.

Table 3. Ten States with the Highest Federal Flood Insurance Claims Payments: 1988-2007

State	Number of Policies Issued (As of 4/30/08)	Total Claims	Total Payments (Nominal $)
Louisiana	501,555	241,807	$14,985,570,820
Florida	2,184,568	122,340	3,311,749,199
Mississippi	78,163	31,738	2,680,787,295
Texas	670,050	91,197	2,423,449,920
Alabama	54,763	21,477	837,190,270
North Carolina	134,509	41,310	736,848,516
New Jersey	226,843	45,838	730,629,078
Pennsylvania	67,311	31,369	673,297,134
New York	148,462	29,993	482,461,366
South Carolina	198,963	15,478	416,677,961

Source: U.S. Department of Homeland Security, Federal Emergency Management Agency.

LESSONS LEARNED FROM PREVIOUS FLOODS

Several flood insurance, risk assessment/mapping and floodplain management regulatory issues were illustrated by the Midwestern floods of 1993 and 2008, and the 2005 hurricanes.

The 1993 Midwest Floods

After the 1993 Midwest floods, the Clinton Administration commissioned a White House study, led by Army Brigadier General Gerald E. Galloway, to determine what could be done to reduce future flood damage. Central to the findings was the labeling of the flood protection system in the upper Mississippi Basin as:

... a loose aggregation of federal, local, and individual levees and reservoirs ... [that] does not ensure the desired reduction in the vulnerability of floodplain activities to damages.[21]

The "Galloway Report" concluded that the 1993 flood was a significant but not unprecedented rainfall-river event, and that such floods would probably occur again. Further, the study noted that although the goals of floodplain management are clear and the means to carry it out existed, improvement and refocusing were needed.[22]

The responsibilities for flood management are largely within state and local governments. Some 15 years later, the nation's levee system remains as an uncoordinated assortment of levees that are owned and maintained by local governments, agencies, and even individuals. Some disaster experts have called for a more uniform approach to managing the levees.

The 2005 Hurricanes

An important lesson learned from the 2005 hurricanes is the importance of identifying levees and assessing the level of protection they offer citizens. Following the floodwall failures in New Orleans and levee overtopping in many parts of coastal

Louisiana in 2005, President Bush signed into law the *Emergency Supplemental Appropriations Act of 2006 for Defense, the Global War on Terror, and Hurricane Recovery*, that included $30 million to develop and implement a national levee inventory and assessment program.[23] In June 2006, the USACE completed the initial inventory survey of federal levees and created a national database that includes the location, number, and condition of these levees.[24] The next step is to include state, local or privately-owned levees in this inventory to improve understanding of the role of levees as a component of the national flood risk.

The 2008 Midwest Floods

The first lesson that can be learned (or relearned) from the 2008 Midwest flood is that the NFIP might not completely internalize the risk of living and investing in the floodplain nor achieve the level of individual participation or reconstruction of older homes originally envisioned. Critics of the NFIP say the program along with levees encourage too many people to locate in areas susceptible to flood damage, and flood victims reliance on federal disaster assistance for uninsured losses. These tendencies, in many ways, negate the original intent of the NFIP, which was to minimize future flood damages and the corresponding need for federal disaster relief.

Second, many property owners affected by the 2008 Midwest floods might have made location and insurance decisions based on inaccurate or incomplete flood maps. This lesson was also learned in the aftermath of Hurricane Katrina. FEMA has consistently sought to communicate to the public the fact that levees do not eliminate the risk of inundation. The residual flood risk generally has not been priced into federal flood insurance policies, funds have not been set aside for catastrophe losses, nor do the premium rates reflect the affect of coastal erosion and climate change on flood risks. These factors were not contemplated or built into the program at inception. Moreover, the NFIP is providing subsidized flood insurance to homeowners who choose not to take advantage of this financial protection. A Rand Corporation study of the NFIP's mandatory purchase requirement nationwide indicated that only about 49% of single family homes in SFHA are covered by flood insurance.[25]

Based on the certification of levees as providing at least protection from the 100-year flood, property owners may not purchase flood insurance, yet they may face significant uninsured losses if the levee fails or is overwhelmed. Many homeowners were told or wrongly concluded they were not at risk because they resided behind a certified levee and, therefore, were outside the federal requirement to purchase flood insurance.

An illustration of the national consequences of not pricing or reserving for residual risk (due to infrastructure failure) might be helpful. The most costly flood in the 40-year history of the NFIP were caused not by rainfall-river flooding but by breeched or overtopped levees protecting the City of New Orleans from coastal storm surges. According to FEMA, some 75-80% of the area behind the levees were designated SFHA (high risk zone) due to rainfall. There was an explicit flood insurance purchase requirement in effect in the affected areas. Still, the NFIP assumed the levees were going to hold back storm surge floods and therefore did not price the policies to reflect the failure or overtopping of levees. The lack of understanding of the national flood risk, the inadequate communication of that risk, and diminished capabilities in flood risk management due to inaccurate or out-of-date flood hazard maps are weaknesses in the program.

Third, the price charged for federal flood insurance could understate the risk because flood hazard data might not be accurately reflected on flood maps and in the underwriting process. In practical terms, the 2008 Midwest floods have exposed the public safety risks associated with levee systems and an over-reliance on levees and other structural flood control measures designed to mitigate future flood losses across the nation. Given the vulnerability of the large number of levees that are not adequately inspected and maintained, there is an increased risk that the NFIP will continue to experience unexpected losses and fiscal deficits, potentially requiring future NFIP borrowing from the U.S. Treasury.

The high degree of uninsured flood losses during the 2008 floods could raise the policy question of who should appropriately bear the cost of the decision to live in potentially high-risk areas, including areas behind levees, dams and other flood control structures. In the absence of flood insurance, the cost of repairing the flood damage will be borne either by the property owner from their own financial resources or through federal disaster assistance — not flood insurance payments.

FINANCIAL STATUS OF NFIP

The NFIP experienced only one catastrophic loss year in its 40 year history, impairing the program's ability to pay current obligations, administrative expenses, as well as interest on the debt to the Treasury. Table 4 shows the NFIP has a current fiscal deficit of $17.4 billion as a result of claims from Hurricanes Katrina and Rita and having to borrow from the Treasury. The 2005 hurricane-related flood claims exceeded the cumulative claims payments since the program's inception.

In an attempt to both protect the NFIP's integrity after the 2005 hurricanes and ensure FEMA has the financial resources to cover its existing commitments, Congress passed, and the President signed into law, legislation to increase the NFIP's borrowing authority to allow the agency to continue to pay flood insurance claims: first to $3.5 billion on September 20, 2005;[26] to $18.5 billion on November 21, 2005;[27] and finally to $20.8 billion on March 23, 2006.[28] FEMA paid $176 million in interest to the Treasury in 2006, $718 million in 2007, and expects to pay $734 million in 2008.

The 2005 floods exposed significant vulnerability in the administration and oversight of the program. It is unlikely that the $17.4 billion treasury debt will be repaid within the next 10 years given annual interest payments of about $1 billion and annual premium income of approximately $2.3 billion.[29] Experts agree that even if FEMA increased flood insurance rates up to the maximum amount allowed by law (10% per year), the program would still not have sufficient funds

Table 4. History of U.S. Treasury Borrowing Under the National Flood Insurance Program (As of July 31, 2008, $ Constant)

Fiscal year	Amount borrowed	Amount repaid	Cumulative debt
Prior to FY1981[a]	$ 917,406,008	$ 0	$ 917,406,088
1981	164,614,526	624,970,099	457,050,435
1982	13,915,000	470,965,435	0
1983	50,000,000	0	50,000,000
1984[b]	200,000,000	36,879,123	213,120,877
1985	0	213,120,877	0
1994[c]	100,000,000	100,000,000	0
1995	265,000,000	0	265,000,000
1996	423,600,000	62,000,000	626,600,000
1997	530,000,000	239,600,000	917,000,000
1998	0	395,000,000	522,000,000
1999	400,000,000	381,000,000	541,000,000
2000	345,000,000	541,000,000	345,000,000
2001	600,000,000	345,000,000	600,000,000
2002	50,000,000	650,000,000	0
2005[d]	300,000,000	75,000,0000	225,000,000
2006	16,660,000,000	0	16,885,000,000
2007	650,000,000	0	17,535,000,000
2008 to date	50,000,000	225,000,000	17,360,000,000
Total	$21,719,535,534	$4,359,535,534	$17,360,000,000

Source: U.S. Department of Homeland Security, Federal Emergency Management Agency's Office of Legislative Affairs.

Note: Borrowings through 1985 were repaid from congressional appropriations. Borrowings since 1994 have been repaid from premium and other income.

a Balance forward from U.S. Department of Housing and Urban Development.
b Figure for the $213.1 million in cumulative debt in 1984 were provided by FEMA. It reflects additional cost outside of the insurance program.
c Of the $100 million borrowed, only $11 million was needed to cover obligations.
d The NFIP borrowed $300 million in 2005 to pay claims from the 2004 hurricane season. Note: Hurricanes Katrina, Rita and Wilma struck in the fall of 2005, after the 2006 fiscal year began.

to cover future obligations for policyholder claims, operating expenses, and interest on debt stemming from the 2005 hurricane season.[30]

By law, the NFIP does not operate under the traditional definition of insurance solvency; rather, the program operates under a congressional mandate of annual limits on premium increases, premium discounts (subsidies) for certain structures in flood-prone areas, and actuarial premium on other structures. Also, unlike private insurers, the NFIP rates are set at levels that make the program self-supporting for the historic average loss year. The program does not generate sufficient premium income to cover flood insurance claims and expenses and build a reserve fund for future catastrophic loss years.

POLICY ISSUES

Recognizing the unprecedented financial and regulatory challenges facing the NFIP, some insurance market analysts and policymakers would maintain that the program's purpose and framework of "carrots and sticks" be re-examined. Changes could occur in several areas: long-term financial solvency, premium structure reform, mandatory purchase requirements, risk assessment and flood hazard mapping, accounting for "write-your-own" companies, and multi-peril coverage.

Long-Term Financial Solvency

A key policy issue for Congress is whether the NFIP should be self-sustaining and how best to pay claims in years with catastrophic losses. According to disaster experts, options for improving the NFIP's financial solvency might include steps to:

- Dramatically reduce the financial cost of multiple flood insurance payments (i.e., repetitive loss properties) that account for a disproportionate share of the NFIP's total claims payouts. This would require eliminating the premium subsidies available to repetitive loss properties.
- Strengthen floodplain management regulations designed to restrict development in high risk areas and require new construction to be elevated three feet above the base flood elevation (BFE).

- Improve flood risk assessment and mapping of the nation's floodplains and include 500-year floodplains and areas behind levees.
- Strengthen and enforce mandatory purchase requirements.
- Forgive the full debt owed by the NFIP to the Treasury.
- Require actuarially-based premiums in the NFIP.

There is no consensus on any of these options. Some aspects of all of them, however, are now being considered in bills — H.R. 3121 and S. 2284 — pending in the 110th Congress.[31]

Premium Subsidies

Federally subsidized flood insurance is offered to encourage participation in the NFIP by communities and the purchase of flood insurance by individuals. Subsidized flood insurance premiums are possible because the government is positioned through loans to the NFIP and otherwise, to spread losses over time in the event of catastrophic flood losses. Owners of properties built prior to the issuance of a community's flood hazard map typically pay rates that are less than full actuarial rates and are exempted from the NFIP's floodplain management standards. The NFIP, however, requires all new and substantially improved buildings to be constructed to or above the elevation of the 1%-annualchance flood. Buildings constructed after December 31, 1974 or after the publication of a flood insurance rate map (FIRM) are charged an actuarial premium that reflects the property's risk of flooding.

Premium subsidies were considered necessary because occupants often did not understand the flood risk when they built in these areas (flood maps were not available); there were no public safeguards prohibiting the occupancy of this land; and subsidies of pre-FIRM structures could provide an incentive to local communities to participate in the program and discourage unwise future floodplains construction.[32] NFIP's premium subsidies were intended to be phased out over time, as the number of pre-FIRM properties (and accompanying subsidies) would gradually diminish as they were damaged and rebuilt/relocated under stronger floodplain management and building codes.

Congress is considering legislative initiatives (H.R. 3121 and S. 2284) that would: (1) eliminate premium subsidies for non-primary residences, commercial properties and repetitive loss properties; (2) increase the allowable annual rate increase in NFIP policies; and (3) create new strata of flood insurance rates to accurately reflect the variations in risk within individual zones.

Solvency and Actuarial Soundness

Congress did not set up the NFIP on an actuarially sound basis when it authorized subsidized rates for pre-FIRM structures without providing annual appropriations to fund the subsidy. In order to make up the subsidized premium shortfall, FEMA established a rating methodology that consisted of a requirement to earn a target level of premium income for the program as a whole that is at least sufficient to cover administrative expenses and losses relative to what FEMA calls the "historical average loss year." The premium level generated to cover the historical average loss year must accommodate the combined effect of the portion of NFIP business paying less than full risk premiums and the portion of the business paying full risk premiums.

Several additional options might be considered to strengthen the financial solvency and actuarial soundness of the NFIP: address the repetitive loss properties (RLPs) problem, create a catastrophe reserve fund for catastrophic loss years, create incentives for private sector participation, and forgive the debt.

Repetitive Flood Loss Properties

Approximately 1% of insured properties, so-called repetitive loss properties, are responsible for approximately 30% of all program claim costs.[33] Efforts are underway to phase out premium subsidies on RLPs through voluntary buyouts or the imposition of full actuarially based rates for RLP owners who refuse to accept FEMA's offer to purchase or mitigate the effect of flood damage. FEMA's Pilot Severe Repetitive Loss Program (SRLP) calls for voluntary buyouts of SRLP and conversion of the land in perpetuity to open space uses.

Encourage Private Sector Participation

One option to address the NFIP's long-term solvency is to undertake efforts to shift flood insurance back to the private insurance markets and open the federal program to competitive bid contractors under the Write-Your-Own program.[34] It might be possible to plan for a higher degree of private sector involvement by requiring private insurers to "make available" private flood insurance policies at actuarially determined prices in flood-prone areas with the federal government providing federal reinsurance that would be self-supporting in the long run. Some economists have suggested that floods and other catastrophic risks may now be insurable because of insurer's ability to transfer risk to the capital markets through securitization, and assess catastrophe modeling and other analytical techniques that permit more accurate pricing of policies.[35]

FEMA has a responsibility to examine the NFIP's contingent liabilities and recommend ways to provide financial stability to the federal flood insurance program. This activity is performed in conjunction with the program's annual rate-setting process. In 2000, FEMA undertook a study with the assistance of accounting firm Deloitte & Touche to explore alternative financing arrangements to reduce the need for U.S. Treasury borrowing. FEMA was concerned about the NFIP's erratic cash flow and the potential for catastrophic losses within a short period of time. The option that received the most attention was to create a special financial reinsurance vehicle to finance catastrophic loss years.[36] After review by the Office of Management and Budget (OMB), this option was not adopted because it was determined that the cost to borrow from the U.S. Treasury was cheaper. A similar option has been suggested to require private insurers to sell flood insurance coverage with a federal reinsurance backstop.

Mandatory Purchase Requirements

Federal flood insurance is mandated for all structures with federally backed loans or mortgages located in SFHA identified in FEMA flood insurance rate maps. To increase the level of total NFIP insurance coverage, some experts have suggested: (1) updating the floodplain maps, especially of areas protected by levees of questionable reliability; (2) increasing the awareness of flood risk, especially where there is a substantial residual risk of catastrophic flooding; (3) requiring mandatory flood insurance in areas where there is a substantial risk of flood; (4) requiring escrow of flood insurance premiums; and (5) requiring additional property owners in residual risk areas to purchase flood insurance, particularly those in the 500-year floodplain behind levees or dams. Debt Forgiveness. The total expected cost to the Treasury to forgive the NFIP's debt was $17.4 billion, as of May 1, 2008. This amount includes $50 million borrowed in early April to meet semi-annual interest payments. The NFIP's outstanding borrowing is the combined result of paying insurance claims and servicing the debt on the borrowing.

Risk Assessment and Flood Hazard Maps

Risk assessment and floodplain mapping are important components of the NFIP's ability to allocate the cost of the program across all policyholders. Flood maps show specific flood zones that correspond to the level of risk and premium rates. In addition to establishing flood insurance rates, flood maps are used to: (1) establish minimum elevation levels for new construction and guide development in floodplains; (2) determine whether a property is located in a SFHA, and, therefore, whether the owner is required to purchase flood insurance in order to secure a federally regulated or insured mortgage; (3) facilitate lender enforcement of mandatory flood insurance purchase requirements; and (4) provide risk information for underwriting and rating applications for flood insurance under the NFIP.

The 2008 Midwest floods has once again raised awareness concerning the accuracy of FEMA's flood hazard maps. A threefold set of problems was revealed. First, the actual insurance cost of living in a floodplain has not been reflected in the cost of ownership. The government generally finds it difficult to assess risk because of political pressure not to differentiate one person or firm from another in the way the private sector would. This difficulty might lead to large and hidden cross-subsidies. Moreover, governments are susceptible to pressures not to enforce certain regulations, particularly after a catastrophic flood event. Second, according to FEMA, approximately 25% of all flood claims have been on properties located outside of currently designated 100-year floodplains. Inaccurate flood maps typically result in unexpected flood damages, uninsured properties and larger than expected federal emergency disaster assistance expenditures. Third, many people continue to underestimate their vulnerability to floods.

Flood Map Modernization

FEMA is currently engaged in a multi-year nationwide Flood Insurance Study (FIS) to revise and update previous FIS/FIRMs into more reliable easy-to-use digital flood insurance rate maps (DFIRMs). FEMA intends to consolidate separately published FIS and FIRMs into one seamless continual FIS report and FIRM that depicts flood hazards nationally. The DFIRMs call for updating FIRMS to a GIS database format that allows ease of modification, electronic access and transmission, and the ability to incorporate more detailed topographic information and use of information across various platforms.

Advisory Flood Recovery Maps

Following the 2005 hurricane season, FEMA issued new advisory base flood elevations (ABFE) for new construction and the rebuilding of structures that were more than 50% destroyed by Hurricanes Katrina and Rita. FEMA indicates that if communities are to be rebuilt after a major flooding event, such as the 2005 hurricanes or the 2008 Midwest flooding, they must be elevated at or above the 1% annual chance flood elevation. According to FEMA, structures built to this standard, as a class, sustain 70% less damage than older buildings. In the case of New Orleans and the surrounding communities, the ABFE requires new construction to be three feet higher than base flood elevation (BFE) for the community's old FIRMs. Some areas that previously were not in a SFHA are now delineated as flood zones. Homeowners in these newly-designated SFHAs would be obligated under federal law to purchase flood insurance.

Residual Risks Behind Levees

The importance of levees in flood risk reduction received much public attention after the levees that protected New Orleans breached and caused massive flooding. In the process of FEMA's development of a countrywide DFIRM, the agency, in coordination with the USACE, must certify levees as providing protection to the 1% annual chance flood elevation. FEMA must have documentation (e.g., maintenance records and engineering reports) from the levee owner that stipulates the levee meets certain standards before it could be shown on a DFIRM as protecting against the 1% annual chance flood. If levees are not certified, FEMA could designate the area behind the levee to be a SFHA which would effectively require homeowners in these areas to purchase flood insurance.

On September 25, 2006, FEMA issued *Revised Procedure Memorandum No. 43 — Guidelines for Identifying Provisional Accredited Levees* to give updated guidance to community officials or other parties seeking accreditation of a levee and the required data and documents to accomplish this task.[37] The memorandum established procedures for provisionally certifying levees in preparation for DFIRMS and the FEMA's map modernization program (MapMod).

FEMA's provisional accredited levee (PAL) procedures were designed to give levee owners more time to gather necessary data and documents needed to prove a levee should be certified. Once FEMA issues an agreement letter that outlines the deficiencies that the levee owner must resolve in order to receive FEMA's levee certification, the community has 90 days to sign and 24 months after signing to submit final documentation. FEMA requires a professional engineer to certify and seal the levee certification. During the 24 month period, the area protected by the levee will be mapped as a shaded zone X (zone outside the Special Flood Hazard Area).

The intent is to regulate flood risk behind levees. The problem is that, although the USACE had successfully inventoried some levees, identified deficiencies in some levees, and communicated information to some levee owners and FEMA, thousands more state, local government, and privately-owned levees have not been similarly identified, evaluated and inventoried. Therefore, FEMA is presumably not prepared for new DFIRM issuance in the MapMod program.

Accountability for Write-Your-Own Companies

Another issue facing the NFIP is the uncertainty about claims adjustment practices when a single event causes both flood damages (NFIP insured) and wind damages (privately insured). As some observed after the 2005 hurricanes, WYO (private) insurers might have an inherent conflict of interest in adjusting NFIP insurance claims and determining whether a loss was attributable to wind (privately covered) or flooding (insured under the NFIP).

The issue in contention is whether FEMA has adequate procedures and collects accurate information on damages to ensure claims paid by the NFIP cover only those damages caused by flooding. Insurers typically adjust flood claims along with their own, creating this internal conflict of interest.[38] To address this issue, Congress is considering whether to: (1) prohibit WYO insurers from including anti-concurrent causation language in their homeowners' policies;[39] (2) establish a Flood Insurance Advocate Office to strengthen the oversight of the WYO program; (3) require FEMA to review and conduct rulemaking on WYO insurer reimbursements so that reimbursements and actual administrative expenses are aligned; and (4) fully implement the 2004 Flood Insurance Reform Act, which some analysts believe would improve communications and assure the proper education and training standards for insurance agents.

Multiple-Peril Coverage for Wind and Flood Damages

As noted earlier, Congress is considering legislation — H.R. 3121 — that would create a combined federal insurance program with coverage for both wind and flood damage. Proponents of adding the wind peril say it is necessary to eliminate coverage disputes when wind and flood both contribute to a loss. Optional wind coverage is also said to be needed because of the difficulty that property owners have in obtaining affordable wind coverage in states along the Gulf and Atlantic coasts. Private insurers have dramatically increased premiums and deductibles, reduced coverage or withdrawn altogether from these areas out of concern about catastrophic risk exposure. In those areas, homeowners must instead purchase their wind coverage from state pools, where the premiums can be prohibitively expensive.

Opponents of adding optional wind coverage to the NFIP believe that there is adequate wind coverage capacity in every state through either the traditional private market or state-sponsored wind pools. They express concern over the

NFIP's ability to properly price an all-perils policy and avoid wide-scale financial deficits in the program following a natural catastrophe.

In addition, the Government Accountability Office (GAO) has cited several concerns about expanding the NFIP to offer wind coverage. They involve: (1) wind hazard prevention standards that communities would have to adopt in order to receive coverage; (2) uncertainty about the adoption of programs to accommodate wind coverage; (3) establishing a new rate-setting process; (4) enforcement of new building codes; and (5) administration and oversight of the program.[40]

Rather than adding an optional wind coverage to the NFIP, the Senate version of H.R. 3121 (S. 2284) takes a different approach to the insurance availability and conflict of interest concerns. S. 2284 would create a bipartisan Commission on National Catastrophe Risk Management and Insurance to study the issue and a National Flood Insurance Advocate, and authorize FEMA to obtain information from WYO insurers about their handling of wind/flood claims.

LEGISLATIVE RESPONSE

In response to the 2008 Midwest floods, Congress has enacted emergency supplemental appropriations legislation to compensate flood victims and is now considering legislative measures — H.R. 3121 and S. 2284 — that would reauthorize and reform the NFIP. On June 30, 2008, President Bush signed into law Public Law 110-252 to, among other things, appropriate $8.48 billion for natural disaster relief and recovery, including $5.64 billion for construction of flood prevention and protection structures in Louisiana and $2.84 billion for flood assistance in Midwestern states. On September 27, 2007, the House passed the Flood Insurance Reform and Modernization Act of 2007, H.R. 3121. On May 13, 2008, the full Senate approved S. 2284, another flood insurance reform bill. The House and Senate have so far not convened a conference committee to reconcile the differences between H.R. 3121 and S. 2284.

Specifically, both H.R. 3121 and S. 2284 would: (1) allow the NFIP to increase premiums by 15% per year, up from 10%; (2) phase out the premium subsidies for second homes, commercial properties, and repetitive-loss properties; (3) increase market penetration by expanding the properties subject to the mandatory flood insurance purchase requirement after areas behind levees are remapped; (4) increase penalties for a lender's non-compliance; and (5) enhance communication to individuals and insurance agents about flood risk.[41]

The bills have some fundamental differences. H.R. 3121 would expand the program to cover damage caused by wind; S. 2284 lacks this provision. S. 2284, but not H.R. 3121, would forgive $17.5 billion debt to the Treasury that the program incurred after the 2005 hurricane season. S. 2284 would create a new Office of Flood Advocate to provide oversight of WYO insurers, as well as a new bipartisan Commission on Natural Catastrophe Risk management and Insurance to study and report to Congress on ways to address the availability of catastrophe insurance coverage in high-risk areas.

End Notes

[1] Following a presidentially declared disaster, individuals and households who have no insurance or are under-insured have recently been provided a variety of federal assistance, including grants for temporary housing and home repairs, low-cost loans to cover uninsured property losses, and other programs to help recover from the effects of the disaster. With respect to assistance to communities, the federal government usually provides 75% reimbursement for disaster-related costs incurred by local and state governments. The remaining 25% in expenses is covered by state and local entities. The intent of federal disaster assistance is to return a damaged region to a functional state, not to pre-disaster conditions. The disaster relief funds distributed by FEMA are from general revenue of the U.S. government. Emergency benefits funded by the federal government are designed to supplement, not to replace, private insurance. Thus, FEMA benefits are intended to be coordinated with insurance coverage.

[2] U.S. Department of Homeland Security, Federal Emergency Management Agency, *Fact Sheet: Hazard Mitigation Grant Program* (Washington: June 2007), p. 1, available online at [http://www.fema.gov/government/grant/hmgp/index].

[3] The acquisition of property after a flood is coordinated at the state and local levels of government. The local community usually identifies potential homes that could be acquired. Federal Emergency Management Agency (FEMA) then provides HMGP monies to the state and local governments to buy the property. Once the property is purchased by the city or jurisdiction, the building is demolished and the land turned into open space in perpetuity. The federal government pays 75% of the acquisition cost, the state 10% and the local community 15%. States receive either 15% or 20% of the total federal disaster assistance given to the state in HMGP funds, depending on whether the state has published an Enhancement Mitigation Plan.

[4] The centerpiece of the nation's floodplain management system has been the National Flood Insurance Program's flood hazard identification and risk mapping, federally based flood insurance, and floodplain management strategies designed to minimize future flood loss and guide development away from flood-prone areas.

[5] Roger A. Pielke, Jr., "Flood Impacts on Society," *Floods*, 2000, vol. 1, p. 133-155.

[6] Roger A. Pielke, Jr. and Mary W. Downton, "Precipitation and Damaging Floods: Trends in the United States, 1932-1997," *2000 American Meteorological Society*, v. 13, p. 3625-3637; located at:[http://sciencepolicy.colorado.edu/admin/publication_files/resource-60-2000.11.pdf]; and, Charles A. Perry, "Significant Floods in the United States During the 20th Century," 2000, *U.S. Geological Survey Fact Sheet 024-00*, p. 4, located at [http://ks.water.usgs.gov/Kansas/pubs/fact-sheets/fs.024-00.html].

[7] U.S. Government Accountability Office, *Climate Change: Financial Risks to Federal and Private Insurers in Coming Decades are Potentially Significant*, GAO Report GAO-07-285 (Washington: March 16, 2007), p 5.

[8] See *Scientific Assessment Captures Effects of a Changing Climate on Extreme Weather Events in North America*, Department of Commerce, National Oceanic and Atmospheric Administration, June 19, 2008, located at [http://www.noaanews.noaa.gov/stories2008/ 20080619_climatereport.html] Also, see National Wildlife Federation, *Heavy Rainfall and Increased Flooding Risk: Global Warming's Wake-Up Call for the Central United States*, located at [http://www.nwf.org/nwfwebadmin/binaryVault/Heavy_Rainfall_and_Increased_Floodin g-Wake-Up_Call_for_Central_U.S2.pdf].

[9] Mississippi River Commission Act of 1879, 46th Cong., 1st sess., June 28, 1879, ch. 43, 21 Stat. 37 (codified as amended at 33 U.S.C §§ 641-653a (2000)).

[10] P.L. 64-367, 39 Stat. 948.

[11] John M. Barry, *Rising Tide: The Great Mississippi Flood of 1927 and How it Changed America*, (New York: Simon and Schuster, 1997).

[12] U.S. Senate, Committee on Banking and Currency, *Insurance and Other Programs for Financial Assistance to Flood Victims: A Report from the Secretary of the Department of Housing and Urban Development to the President, as Required by the Southeast Hurricane Disaster Relief Act of 1865 (Public Law 89-339, 89th Congress, H.R. 11539, November 8, 1965)*, 89th Congress, 2nd sess., September 1966 (Washington: GPO, 1966).

[13] P.L. 90-448, 82 Stat. 573.

[14] Norbert Schwartz, "FEMA and Mitigation: Ten Years After the 1993 Midwest Flood," *Journal of Contemporary Water Resource & Education*, Issue 130, March 2005, p. 36. [15] *White House Floodplain Management Task Force*, "Sharing the Challenge: Floodplain Management Into the 21st Century: Report of the Interagency Floodplain Management Review Committee to the Administration Floodplain Management Task Force," June 1994, located at [http://www.floods p. 131.

[16] P.L. 93-234; 87 Stat. 975.

[17] P.L. 93-234; 87 Stat. 975.

[18] P.L. 103-325; 108 Stat. 2255.

[19] P.L. 108-264; 118 Stat. 712.

[20] For a more detailed list of challenges facing the NFIP, see, U.S. General Accounting Office, *Challenges Facing the National Flood Insurance Program*, GAO Report GAO-03-606T (Washington: April 1, 2003).

[21] White House Floodplain Management Task Force, *Sharing the Challenge: Floodplain Management Into the 21st Century: Report of the Interagency Floodplain Management Review Committee to the Administration Floodplain Management Task Force*, June 1994, located at [http://www.floods p. 131.

[22] Ibid.

[23] P.L. 109-234, 120 Stat. 455.

[24] See, U.S. Army Corps of Engineers, *Fact Sheet: National Levee Safety Program*, located at [http://www.hq.usace.army.mil/cepa/releases/leveesafetyfactsheet.pdf].

[25] Rand Institute for Civil Justice, *The National Flood Insurance Program's Market Penetration Rate: Estimates and Policy Implications* [http://www.rand.org/pubs/technical_reports/2006/RAND_TR300.pdf].

[26] P.L. 109-65; 119 Stat. 1998.

[27] P.L. 109-106; 119 Stat. 2288.

[28] P.L. 109-208; 120 Stat. 317.

[29] See Letter from Donald B. Maroon, Acting Director of Congressional Budget Office, to Honorable Judd Gregg, Chairman, Committee on the Budget, March 31, 2006, located at [http://www.cbo.gov/ftpdocs/72xx/doc7233/05-31-NFIPLetterGregg.pdf].

[30] Ibid.

[31] H.R. 3121, Flood Insurance Reform and Modernization Act of 2007, was introduced by Rep. Waters. S. 2284, with the same title as H.R. 3121, was introduced by Sen. Dodd. See CRS Report RL34367, *Side-by-Side Comparison of Flood Insurance Reform Legislation in the 110th Congress*, by Rawle O. King.

[32] PriceWaterhouseCoopers, *Study of the Economic Effects of Charging Actuarially Based Premium Rates for Pre-FIRM Structures,* Washington, May 14, 1999, p. 1-2.

[33] U.S. Government Accountability Office, *National Flood Insurance Program: Actions to Address Repetitive Loss Properties,* GAO Report GAO — 04-401T (Washington: March 25, 2004), p. 6.

[34] The idea is to outsource the marketing, underwriting, and policy administration of the NFIP through a single or a few private insurers under a contractual arrangement. Currently, any insurer wishing to participate in the NFIP's Write-Your-Own program can do so.

[35] Dwight M. Jaffee and Thomas Russell, "Can Securities Markets Save the Private Catastrophe Insurance Market?" paper delivered at eh Asian-Pacific Risk and Insurance Association Conference, July 19, 1998, p. 11.

[36] Federal Emergency Management Agency, *National Flood Insurance Program: Discussion of Financial Stabilization Possibilities,* FEMA Unpublished Internal Document, November 20, 2000.

[37] See, U.S. Department of Homeland Security, Federal Emergency Management Agency, *Revised Procedure Memorandum No. 43 — Guidelines for Identifying Provisional Accredited Levees,* located at [http://www.fema.gov/library/viewRecord.do?id=2511].

[38] U.S. Government Accountability Office, *National Flood Insurance Program: Greater Transparency and Oversight of Wind and Flood Damage Determinations are Needed,* GAO Report GAO-08-28 (Washington: December 28. 2007), p. 21.

[39] Concurrent causation holds that if two causes (e.g., water and wind) combine to produce a loss or damage, and one of the two causes is excluded, the loss will be covered absent policy wording to the contrary. The "anti-concurrent causation" doctrine was designed to prevent the theory of "concurrent causation" from broadening coverage under standard property insurance policies.

[40] U.S. Government Accountability Office, *Natural Catastrophe Insurance: Analysis of a Proposed Combined Federal Flood and Wind Insurance Program,* GAO Report GAO-08-504 (Washington: April 25, 2008), p. 2.

[41] CRS Report RL34367, *Side-by-Side Comparison of Flood Insurance Reform Legislation in the 110th Congress,* by Rawle O. King.

In: Federal Flood Policy
Editor: James E. Rysanek

ISBN: 978-1-61324-017-5
© 2011 Nova Science Publishers, Inc.

Chapter 5

MIDWEST FLOODS OF 2008: POTENTIAL IMPACT ON AGRICULTURE[*]

Randy Schnepf

SUMMARY

Unusually cool, wet spring weather followed by widespread June flooding across much of the Corn Belt cast considerable uncertainty over 2008 U.S. corn and soybean production prospects. As much as 5 million acres of crop production were initially thought to be either lost entirely or subject to significant yield reductions. Estimates of flood-related crop damage varied widely due, in part, to a lack of reliable information about the extent of plant recovery or replanting in the flooded areas. These circumstances generated considerable market angst and U.S. agricultural prices for corn and soybeans, as reported on the major commodity exchanges, hit record highs in late June and early July. Since then, most of the Corn Belt has experienced nearly ideal growing conditions suggesting the potential for substantial crop recovery, and market prices have weakened accordingly.

On August 12, 2008, USDA released the first crop production estimates for corn and soybeans that have incorporated survey data from the flood-affected regions. According to USDA, U.S. farmers will produce the second largest corn crop on record — 12.3 billion bushels — in 2008, up about 5% from the previous month's forecast, but down over 6% from last year's record

---

[*] This is an edited, reformatted and augmented version of a Congressional Research Services publication, dated August 18, 2008.

crop. USDA's soybean crop forecast of nearly 3 billion bushels is unchanged from July, but up 15% from 2007. These production forecasts reflect three factors. First, flood-related acreage losses appear to be substantially less than initially projected. Second, nearly ideal growing conditions that have persisted across the Corn Belt since late June have contributed to sharp increases in USDA's yield outlook for corn, thus, offsetting flood-related area losses. Third, despite a 17.6% increase in planted acreage in 2008, soybean production is flat due to a diminished yield outlook — largely the result of the lateness of the crop's planting and development, as well as dry conditions in the Delta, the Southeast, and the Northern Plains.

Congress has appropriated nearly $480 million in emergency USDA funding, primarily for conservation activities in flood-affected regions, as part of the FY2008 Supplemental Appropriations Act (P.L. 110-252). USDA has also committed resources to the flood-affected areas including rescue and clean up, food assistance, housing, community assistance, business assistance, and farmer and rancher assistance. In addition, USDA announced permission, on July 7, 2008, to use CRP land for grazing only in disaster and contiguous counties.

In light of current market uncertainties surrounding the 2008/09 supply and demand balance for corn and soybeans, and the outlook for extremely tight supplies by late summer, commodity market prices are likely to remain volatile through the remainder of the growing season. If crop production ultimately proves less than forecast (to be determined at harvest time), it will likely contribute to higher commodity prices, thereby adding to pressure on policymakers over concerns about consumer food price inflation, international food aid availability, and the soundness of policy that dedicates commercial agricultural crops to biofuels production, particularly corn used for ethanol.

BACKGROUND

The United States plays a critical role in global markets for both feed grains and oilseeds. The United States is the world's leading producer and exporter of both corn and soybeans. In 2007 the United States had 42% and 63% shares, respectively, of world corn production and trade, and 32% and 41% shares of world soybean production and trade. As a result of this dominant role, unexpected changes in U.S. production for either corn or soybeans, such as those stemming from the Midwest floods of 2008, can have a major impact on both U.S. and global commodity markets.

During the first half of 2008, U.S. and world agricultural markets for most grains and oilseeds experienced tight supplies and record high prices.[1] The high prices provided a tantalizing incentive for U.S. farmers as they prepared to plant their crops this past spring. In contrast, the dramatic, unexpectedly sharp price increases of the past year have raised costs for livestock feeders and agricultural processors, evoked considerable concern about consumer food-price inflation and international food aid availability, and sparked a global debate — referred to as the "food versus fuel" debate — about the increasing policy trend of dedicating commercial agricultural crops to biofuels production, particularly corn used for ethanol.

Against this backdrop of producer anticipation and consumer angst, substantial new concerns emerged by late June about potential weather- and flood-related production losses to this year's U.S. corn and soybean crops. Widespread, good growing conditions have persisted since the floods adding to the uncertainty over crop production prospects.

U.S. Corn Belt

The Corn Belt is a 13-state region located in the Midwest where corn is the predominant cash crop (Figure 1). It stretches from Ohio through Indiana, Illinois, Iowa, northern Missouri, southern Wisconsin, and Minnesota to the eastern fringe of the Great Plains states of North and South Dakota, Nebraska, and Kansas. The Corn Belt also includes parts of Michigan and Kentucky. Since 2000, these 13 states have accounted for 89% of U.S. corn production (Table 3). Iowa and Illinois, in the heart of the Corn Belt, are the two leading corn-producing states with a combined production share of 36%. Similarly, 88% of U.S. soybean production occurs in the 13 Corn Belt states, with Iowa and Illinois again the two leading producers with a combined share of 32% (Table 4).

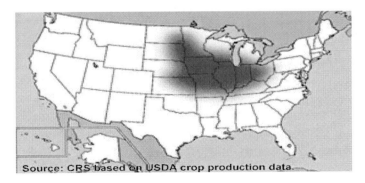

Figure 1. Corn Belt.

USDA'S CURRENT CROP OUTLOOK FOR CORN AND SOYBEANS

On August 12, 2008, USDA released the first survey-based forecast of corn and soybean production for 2008.[2] According to USDA's forecast, U.S. farmers will produce the second largest corn crop on record — 12.3 billion bushels — up about 5% from the previous month's forecast, but down over 6% from the 2007 record crop. USDA's soybean production forecast of nearly 3 billion bushels is unchanged from the July forecast, but up 15% from 2007.

These production forecasts reflect three factors. First, flood-related acreage losses appear to be substantially less than initially projected. Second, nearly ideal growing conditions that have persisted across the Corn Belt since late June have contributed to sharp increases in USDA's yield outlook for corn, thus, offsetting flood-related area losses. Third, despite a 17.6% increase in planted acreage in 2008, soybean production is flat due to a diminished yield outlook — largely due to the lateness of the crop's planting and development, as well as dry conditions in the Delta, the Southeast, and the Northern Plains.

USDA's August crop production forecast appear to have calmed much of the market concern regarding crop losses due to flooding. However, a large portion of the 2008 corn and soybean crops were planted late and, as of early August, remain substantially behind historical development rates.[3] As of August 11, USDA estimates that 30% of corn had reached the dough stage of development compared with the 5-year average of 50%, while only 6% had dented compared with an average of 16% the past five years. Similarly, 60% of soybean plants had set pods compared with the 5-year average of 75%. As a result, market analysts suggest that weather problems could still emerge — such as an early freeze — that could lower yield and production prospects, especially in the more northerly regions where crop development remains behind normal.

USDA Re-Surveys Flooded Areas

USDA's August crop production forecasts reflect growing conditions as of August 1, and incorporate survey data from the flood-affected regions. The yield estimates are based on objective field surveys while the planted and harvested acreage estimates are usually drawn from the June *Acreage* report.[4] However, most of the survey data for the *Acreage* report was collected during the first two weeks of June prior to the worst flooding. In response to the changed circumstances, USDA conducted an extensive re-interview of producers' harvesting intentions in mid-July, in the flood-affected areas of Illinois, Indiana, Iowa, Minnesota, Missouri, and Wisconsin, to supplement the earlier survey data in deriving estimates of abandoned and harvested acres.[5] USDA stated that under a return to normal weather conditions, by mid-July most flooded fields would be dry and affected farmers would be better able to assess their options. Data obtained from the mid-July re-interviews were incorporated into USDA's August 12, 2008, *Crop Production* and *WASDE* reports.

Outlook for Corn Harvested Acres

USDA estimates 2008 U.S. planted and harvested corn area of 86.977 and 79.290 million acres, respectively.[6] This compares with the June *Acreage* estimates of 87.327 million and 78.940 million acres (Table 1). Thus, planted corn acreage has been revised downward 350,000 acres, while harvested acreage was raised by 350,000 acres. Planted area losses occurred primarily in the flood-affected states. Harvested area gains occurred primarily in states outside of the flood regions and is reflected in below-average abandonment rates. High market prices appear to be encouraging farmers to make every effort to harvest more marginal areas that are traditionally abandoned or grazed off by livestock.

Outlook for Soybean Harvested Acres

USDA estimates 2008 U.S. planted and harvested soybean area of 74.783 million and 73.341 million acres, respectively.[7] This compares with the June *Acreage* estimates of 74.533 and 72.121 million acres (Table 2). Thus, planted soybean acreage has been revised upward 250,000 acres, while harvested acreage was raised by 1,220,000 acres. In contrast to corn, soybean harvested area gains occurred primarily in the flood-affected states.

Outlook for Corn Yield

USDA's August estimate of 2008 corn yields was 155 bushels per acre. If realized, this would be the second largest on record behind the 160.4 bushels per acre achieved in 2004. Clearly, excellent weather since late June has boosted the yield outlook. Just a month earlier, in July, USDA had forecast national average corn yields at 148.4 bu./ac. due to the combined effects of slow planting progress, unusually slow plant emergence, and the flooding.[8] Final yields may still vary based on growing conditions through the remainder of the growing season. USDA updates its crop production and market supply and demand estimates monthly.[9]

Outlook for Soybean Yield

USDA's August estimate of 2008 soybean yields was 40.5 bushels per acre. If realized, this would be down nearly 2% from last year's 41.2 and the lowest since 2003. The soybean crop's late development and dryness throughout much of the Southeast, Delta, and Northern Plains appears to be taking its toll. Just a month earlier, in July, USDA had forecast national average soybean yields at 41.6 bu./ac. based on 1989-2007 regional trend analysis adjusted for late planting and emergence.[10] As with corn, final soybean yields may still vary based on growing conditions through the remainder of the growing season.

ESTIMATING CROP LOSSES FOR 2008

Flood-related crop damage assessments generally are made by county and state officials in the affected regions. However, a rough approximation of flood-damaged acres can be obtained by comparing the implied state-level abandonment rates from USDA's August forecasts with the recent eight-year average abandonment rates. If one attributes any change from the 8-year average entirely to the flood, then the data suggest that about 889,000 acres planted to corn and intended for harvest were lost in Iowa (453,000), Illinois (236,000), Indiana (102,000), and Missouri (99,000) — see Table 1.

This "lost" area estimate represents about 1% of the 87.0 million acres planted to corn in 2008. However, projected below-average abandonment rates throughout the remainder of the Corn Belt, particularly in Nebraska, Kansas, South Dakota, Ohio, and in lower-yielding non-Corn Belt states more than offset the lost acres. Applying USDA August yield forecasts to the area-loss calculations suggests that the four major flood-affected states of Iowa (77.5 million bushels), Illinois (40.6), Indiana (15.8), and Missouri (14.4), cumulatively account for 148.2 million bushels of "potentially" lost production.[11] This "lost" production estimate represents 1.2% of the 12.3 billion bushel crop estimate announced by USDA.

Table 1. Estimated Corn Acres Lost Due to June 2008 Floods Based on Predicted Abandonment Rates

State	March[a] Planted	June[b] Planted	June[b] Harvested	August 12, 2008[c] Planted	August 12, 2008[c] Harvested	Abandonment June 2008	Abandonment Aug. 2008	Abandonment Ave: 2000-07	August Implied Area Loss[d]
		1,000 acres				Percent			1,000 acres
Iowa	13,200	13,700	12,280	13,700	12,900	6.6%	5.8%	2.5%	(453)
Illinois	12,600	12,300	11,500	12,200	11,800	6.5%	3.3%	1.3%	(236)
Nebraska	8,800	9,000	8,750	9,000	8,750	2.8%	2.8%	5.0%	198
Minnesota	7,600	7,800	7,250	7,800	7,250	7.1%	7.1%	7.0%	(0)
Indiana	5,700	5,700	5,350	5,600	5,350	6.1%	4.5%	2.6%	(102)
Ohio	4,650	4,650	4,200	4,650	4,200	9.7%	9.7%	15.3%	260
South Dakota	3,900	4,100	3,900	4,100	3,900	4.9%	4.9%	10.0%	209
Kansas	3,650	3,800	3,100	3,750	2,950	18.4%	21.3%	22.9%	57
Wisconsin	3,350	3,350	3,150	3,350	3,150	6.0%	6.0%	6.8%	28
Missouri	3,100	2,900	2,500	2,800	2,600	13.8%	7.1%	3.6%	(99)
Michigan	2,250	2,400	2,150	2,400	2,150	10.4%	10.4%	18.1%	185
Kentucky	2,350	2,350	2,080	2,350	2,080	11.5%	11.5%	11.4%	(1)
North Dakota	1,230	1,230	1,150	1,230	1,150	6.5%	6.5%	7.0%	6

Table 1. (Continued).

State	March[a] Planted	June[b] Planted	June[b] Harvested	August 12, 2008[c] Planted	August 12, 2008[c] Harvested	Abandonment June 2008	Abandonment Aug. 2008	Abandonment Ave: 2000-07	August Implied Area Loss[d]
Corn Belt	72,380	73,280	67,880	72,930	68,230	6.6%	5.7%	3.4%	(45)
Non-Corn Belt	13,634	14,047	11,060	14,047	11,060	21.3%	21.3%	25.3%	569
United States	86,014	87,327	78,940	86,977	13,700	9.6%	8.8%	9.3%	369

Source: NASS, USDA.

a Prospective Plantings, NASS, USDA, March 31, 2008.
b Acreage, NASS, USDA, June 30.
c Crop Production, NASS, USDA, August 12, 2008.
d Calculations are by CRS based on departure from average abandonment rates.

Table 2. Estimated Soybean Acres Lost Due to June 2008 Floods Based on Predicted Abandonment Rates

State	March[a] Planted	June[b] Planted	June[b] Harvested	August 12, 2008[c] Planted	August 12, 2008[c] Harvested	Abandonment June 2008	Abandonment Aug. 2008	Abandonment Ave: 2000-07	August Implied Area Loss[d]
	1,000 acres					Percent			1,000 acres
Iowa	9,800	9,400	8,950	9,500	9,300	2.1%	4.8%	0.5%	(156)
Illinois	8,800	9,100	8,600	9,100	8,950	1.6%	5.5%	0.5%	(102)
Minnesota	7,100	7,100	6,950	7,100	6,950	2.1%	2.1%	1.7%	(32)
Indiana	5,500	5,500	5,200	5,600	5,550	0.9%	5.5%	0.5%	(21)

	March[a] Planted	June[b] Planted	June[b] Harvested	August 12, 2008[c] Planted	August 12, 2008[c] Harvested	Abandonment June 2008	Abandonment Aug. 2008	Abandonment Ave: 2000-07	August Implied Area Loss[d]
State							Percent		1,000 acres
	1,000 acres	1,000 acres	1,000 acres	1,000 acres	1,000 acres				
Missouri	5,200	5,300	5,000	5,300	5,100	3.8%	5.7%	1.2%	(137)
Nebraska	5,000	4,750	4,700	4,750	4,700	1.1%	1.1%	1.2%	8
Ohio	4,500	4,600	4,580	4,600	4,580	0.4%	0.4%	0.5%	3
South Dakota	4,100	4,100	4,040	4,100	4,040	1.5%	1.5%	1.4%	(1)
North Dakota	3,550	3,400	3,340	3,400	3,340	1.8%	1.8%	2.3%	20
Kansas	3,200	3,200	3,100	3,200	3,100	3.1%	3.1%	5.1%	64
Michigan	2,000	1,900	1,890	1,900	1,890	0.5%	0.5%	0.7%	3
Wisconsin	1,650	1,650	1,560	1,700	1,630	4.1%	5.5%	2.1%	(35)
Kentucky	1,330	1,330	1,320	1,330	1,320	0.8%	0.8%	1.2%	6
Corn Belt	**61,730**	**61,330**	**59,230**	**61,580**	**60,450**	**1.8%**	**3.4%**	**1.2%**	**(411)**
Non-Corn Belt	13,063	13,203	12,891	13,203	12,891	2.4%	2.4%	3.8%	184
United States	**74,793**	**74,533**	**72,121**	**74,783**	**73,341**	**1.9%**	**3.2%**	**1.6%**	**(271)**

Source: NASS, USDA

a Prospective Plantings, NASS, USDA, March 31, 2008.
b Acreage, NASS, USDA, June 30.
c Crop Production, NASS, USDA, August 12, 2008.
d Calculations are by CRS based on departure from average abandonment rates.

Table 3. Corn Area, Yield, and Production, U.S. and Corn Belt, Averages for 2000-2007

State	Major Crops[a] Total Planted Area	Acreage Planted	Acreage Harvested	Abandonment rate	Corn Yield	Corn Production	Corn Ave. FarmPrice	Corn Value of Production
	1,000 acres	1,000 acres	1,000 acres	%	bu./ac.	Million	$/bu.	$ Million
Iowa	24,658	12,600	12,281	2.5	162.6	2,002	2.40	4,906
Illinois	23,337	11,606	11,450	1.3	157.9	1,812	2.51	4,656
Nebraska	18,927	8,419	8,000	5.0	147.4	1,182	2.44	2,942
Minnesota	19,764	7,363	6,844	7.0	152.3	1,043	2.33	2,461
Indiana	12,340	5,763	5,610	2.6	150.4	845	2.50	2,138
Ohio	10,201	3,413	3,180	6.8	142.5	454	2.49	1,142
South Dakota	17,103	4,444	3,765	15.3	111.8	425	2.27	975
Kansas	23,045	3,381	3,044	10.0	129.1	395	2.54	1,016
Wisconsin	8,039	3,675	2,835	22.9	135.6	385	2.43	949
Missouri	13,856	2,931	2,825	3.6	130.3	369	2.47	921
Michigan	6,525	2,275	2,015	11.4	127.8	257	2.44	638
Kentucky	5,575	1,236	1,150	7.0	134.0	154	2.62	406
North Dakota	21,578	1,511	1,238	18.1	114.3	142	2.28	355
Corn Belt	204,946	68,616	64,236	6.4	147.0	9,467	2.44	23,506
Non-Corn Belt	117,844	12,307	9,191	33.6	126.1	1,159	2.75	3,182
United States	322,790	80,923	73,428	10.2	144.4	10,625	2.46	26,688

Source: National Agricultural Statistics Service, USDA, Online Agricultural Statistics Database, July 9, 2008. Note: States are ranked by average production for the six-year period.

a. USDA defines major crops as barley, corn, cotton, millet, oats, peanuts, rapeseed, sunflower, rice, rye, sorghum, and wheat.

Table 4. Soybean Area, Yield, and Production, U.S. and Corn Belt, Averages for 2000-2007

State	Major Crops[a] Total Planted Area	Soybeans Acreage Planted	Harvested	Abandonment	Yield	Production	Price	Value of
	1,000 acres	1,000 acres		%	bu./ac.	Million bu.	$/bu.	$ Million
Iowa	24,658	10,213	10,165	1.4	46.4	470	6.36	2,937
Illinois	23,337	9,981	9,929	0.3	44.6	442	6.45	2,777
Minnesota	19,764	7,138	7,019	0.8	39.6	277	6.15	1,681
Indiana	12,340	5,463	5,434	0.9	46.3	252	6.34	1,558
Nebraska	18,927	4,650	4,593	1.5	44.9	206	6.02	1,234
Ohio	10,201	4,481	4,459	1.3	42.4	189	6.24	1,181
Missouri	13,856	4,981	4,923	1.1	36.7	181	6.27	1,119
South Dakota	17,103	4,075	4,016	0.8	33.8	135	5.94	791
North Dakota	21,578	2,940	2,871	0.8	31.6	90	5.89	545
Kansas	23,045	2,825	2,680	0.5	30.1	82	6.21	505
Michigan	6,525	2,000	1,983	5.9	36.6	72	6.19	445
Wisconsin	8,039	1,578	1,545	23.0	38.8	60	6.04	355
Kentucky	5,575	1,253	1,238	1.1	39.1	49	6.43	303
Corn Belt	204,946	61,576	60,857	1.2	41.2	2,503	6.24	15,403
Non-Corn Belt	117,844	11,185	10,767	3.9	32.4	349	6.16	2,153
United States	322,790	72,763	71,623	1.6	39.8	2,852	6.25	17,584

Source: National Agricultural Statistics Service, USDA, Online Agricultural Statistics Database, July 9, 2008. Note: States are ranked by average production for the six-year period.

a. USDA defines major crops as barley, corn, cotton, millet, oats, peanuts, rapeseed, sunflower, rice, rye, sorghum, and wheat.

Applying the same abandonment rate methodology to soybeans suggests that projected area loss related to bad weather and flooding amounts to nearly 400,000 acres in the Corn Belt, partially offset by 184,000 acres of below-normal abandonment in non-Corn Belt states (Table 2). This "lost" area estimate represents about 0.2% of the 74.8 million acres planted to soybeans in 2008. Applying USDA August yield forecasts to the area-loss calculations suggests that for soybeans there are six major flood-affected states that cumulatively account for 63 million bushels of "potentially" lost soybean production: Iowa (19.1 million bushels), Illinois (19.0), Indiana (12.5), Missouri (8.8) Wisconsin (2.4), and Minnesota (1.3).[12] This "lost" production estimate represents 2.1% of the estimated 3 billion bushel crop.

UNUSUAL SPRING WEATHER ACROSS THE U.S. CORN BELT

Wet, Cool Weather Persists Since Late 2007

The 2008 Midwest weather-related crop problems — the late planting start, slow crop development, and severe June flooding — were precipitated in 2007 by above-normal rainfall and a cold, wet winter that saturated soils. In Iowa, 2007 was the fourth-wettest year on record.[13] The unusually cool, wet conditions persisted through spring 2008. Again citing Iowa, which was subsequently hit the hardest by June floods (Figure 2), as an example, the first six months of 2008 represented the wettest January-to-June period on record. Cool weather inhibited evaporation rates, thus slowing the soil's rate of drying. As a result, many regions of the Corn Belt were saturated and vulnerable to erosion, ponding (standing water), and flooding when heavy storms in late May and early June dropped additional rainfall.

Planting Date Is Critical for Optimal Yields

Traditionally, farmers plant corn as early as possible because early planting provides the greatest potential to achieve maximum yields.[14] Corn is usually planted ahead of soybeans. Early corn planting is discouraged by wet or cold soils (below 50° F). As a result, more southerly regions tend to have earlier optimal planting dates. In Iowa the optimal corn planting dates are between April 20 and May 5. Yields begin to drop off as the planting date is delayed. A significant yield reduction occurs when the planting date is extended to late May or June.

Similarly, the optimal soybean planting date in Iowa is the last week of April for the southern two-thirds of the state, and the first week of May for the northern third. Optimal planting dates in more northerly latitudes, such as in Minnesota or Wisconsin, occur slightly later and have a smaller window for delayed planting.

This year's excessive rainfall coupled with unusually persistent cold ground temperatures delayed both corn plantings and subsequent plant emergence across much of the prime growing region of the Corn Belt. By May 11, only 51% of intended corn area in the Corn Belt had been planted compared with the previous 5-year average (2002-07) of 77%.[15] Similarly, only 11% of intended soybean area had been planted compared with the 5-year average of 29%. The late start pushed key plant development stages of the corn growth cycle into the hotter weeks of July and August, when it is susceptible to heat stress and dryness, and later into the fall, when the possibility of an early freeze can prematurely end ear or pod filling. In addition, a late start to corn generally implies a late start to soybean production (whose planting generally follows corn), with similar growth concerns.

By May 27, 88% of intended corn acres had been planted versus the 5-year average of 94%, and 52% of intended soybean acres versus 5-year average of 67%. However, equally if not more critical were the on-going delays in plant emergence for both crops. Only 52% of planted corn had emerged compared with a 5-year average of 76% emergence, and only 12% of planted soybeans had emerged versus the 5-year average of 34%. As a result, crop yield concerns were already developing by late May.

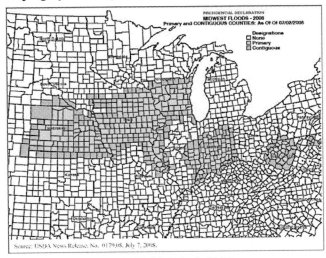

Source: USDA News Release, No. 0179.08, July 7, 2008.

Figure 2. Counties Designated as Presidential Disaster Areas.

June Flooding Ravages Key Growing Areas

With soils already saturated and yield concerns mounting, widespread, heavy rains across the Corn Belt in late May and early June washed out substantial areas recently planted to crops. In addition, they produced severe erosion and gullying, and left saturated soils and standing water in many fields. But most damagingly, the rains triggered widespread flooding across the heart of the Corn Belt. Thousands of acres of prime cropland in Iowa, Nebraska, Illinois, Indiana, Wisconsin, and Missouri were flooded by rivers that swelled their banks and caused levees to break as the storm surge moved through the Mississippi River watershed. Indiana's agriculture director said that the June floods had likely caused the worst agriculture disaster in the state's history, damaging nearly 10% of corn and soybean crops.[16]

The flooding likely led to the abandonment of substantial planted crop acreage, and to yield losses in those crops that survived the flooding but were subject to extended periods of standing water or waterlogged soil.[17] A further concern of saturated soils persisting during the early stages of plant development (particularly for late-planted crops) is that corn plants are more likely to develop shallow root systems, which, in turn, increase their vulnerability to heat and dryness later in the growing cycle.

Initial attempts to ascertain the extent of the crop damage were difficult because the eventual yield and production outcomes for the affected areas depends on how quickly flood waters recede and whether plant growth resumes or new seed is planted. For many farmers, by late June the replanting window for corn had already closed or was approaching faster than the soils were drying. In many cases, the indemnities offered under federally subsidized crop insurance represented greater potential remuneration than incurring the costs of replanting subject to a substantial reduction in yield coverage (due to the late planting date). Replanting to soybeans was an option for some, but many farmers who initially planted corn had already applied a round of herbicide, which would likely damage or kill the soybean seed.

FLOOD-RELATED CROP PRODUCTION AND MARKETING ISSUES

Transportation Infrastructure Damage

While spring flooding in the upper Midwest had caused problems for barge traffic earlier in the year, the extreme rain in June stopped navigation on a nearly 300-mile stretch of the Mississippi River.[18] Major parts of the rail network in the Midwest were damaged, and several major highways in Iowa were temporarily closed. The transportation infrastructure damage resulted in significant delays as grain shipments were rerouted and repairs were underway. By July 6, the Mississippi River had re-opened to commercial traffic, but substantial delays persisted. As a result, many shipments of corn and soybeans were still being rerouted to Texas Gulf ports.

Agricultural Processing and Storage Facilities Disruptions

The flood waters partially submerged many grain elevators and storage facilities, as well as two ethanol plants in Iowa. However, the main damage to agricultural marketing and processing facilities located in the flood-affected region was economic and primarily attributable to delays in the arrival of primary commodity shipments due to the transportation infrastructure damage. Many grain elevators, ethanol plants, soybean crushing plants, and other agricultural processing facilities were temporarily closed or operating at reduced capacity in the weeks immediately following the floods. The Iowa Renewable Fuels Association initially estimated that more than 300 million gallons (annualized) of ethanol production capacity were off line on June 13.[19] In addition, several grain elevators and other types of storage facilities located within the flood zone were damaged. The number of grain elevators damaged and the potential volume of corn and soybean stocks lost is not yet available but is being evaluated by USDA.

Livestock Losses and Disposal Issues

The suddenness of the floods across eastern Iowa resulted in the deaths of possibly thousands of head of livestock, particularly hogs. However, preliminary assessments for the state of Iowa suggest that the actual livestock mortality tally may be substantially lower than initially feared.[20] It appears that most producers had sufficient advance warning of potential flood conditions to move their animals to a safer location ahead of the floods.

THE FEDERAL RESPONSE

Designated Disaster Areas

The President is authorized — by the Robert T. Stafford Disaster Relief and Emergency Assistance Act (the Stafford Act) — to issue major disaster or emergency declarations in response to catastrophes that overwhelm state and local governments.[21] Iowa, with 85 of its 99 counties declared eligible for either individual or public a federal disaster area, appeared to be the hardest hit by the storms and flooding.[22] However, counties in Indiana (44 counties), Illinois (24), Minnesota (4), Wisconsin (30), Nebraska (53), as well as West Virginia (18), were also identified as primary disaster areas related to the spring floods (Figure 2).[23]

A Presidential declaration results in the distribution of a wide range of federal aid to individuals and families, certain nonprofit organizations, and public agencies in the designated areas. Congress appropriates money to the Disaster Relief Fund (DRF) for disaster assistance authorized by the Stafford Act, which is administered by the Federal Emergency Management Agency (FEMA) within the Department of Homeland Security (DHS). Appropriations to the DRF remain available until expended. However, DRF funds are not available to cover agricultural production losses. Instead, USDA offers several permanently authorized programs to help farmers recover financially from a natural disaster, including federal crop insurance, the non-insured assistance program (NAP), and emergency disaster loans.[24]

Agricultural Assistance

USDA is actively engaged in committing resources to the flood response. In this regard, USDA has undertaken a broad range of activities in the flood-affected areas including rescue and clean up, food assistance, housing, community assistance, business assistance, and farmer and rancher assistance.[25]

Congress has appropriated nearly $480 million in emergency USDA funding specifically targeted to 2008 Midwest flood response activities as part of the FY2008 Supplemental Appropriations Act (P.L. 110-252). This funding is available for eligible farmers to defray the cost of clean-up and rehabilitation of farmland and watersheds following a disaster.[26] Of the total amount available, $89.4 million is for the Emergency Conservation Program, which assists farmers in the cleanup and restoration of farmland damaged by a natural disaster, and $390.5 million is for the Emergency Watershed Protection Program, which is designed to relieve imminent hazards created by natural disasters and to alleviate future flood risk.

The 2008 farm bill (P.L. 110-246) included provisions that authorized and funded a new four-year supplemental revenue crop disaster program (for crop years 2008-2011).[27] However, without advance payments, no emergency supplemental disaster assistance for 2008 crop and livestock losses will be available before October 2009. This is because — according to the farm bill disaster program's design — the payment formula used to determine the level of payments for 2008 crop and revenue losses is based on national average market prices which will not be known until Fall 2009. USDA claims that it does not have the authority to make advance payments. Some policymakers want to amend the farm bill to require USDA to make advance payments, while several farm groups contend that USDA already has the flexibility and should exercise its authority.

USDA has also been under considerable pressure from Members of Congress and groups representing the livestock, biofuels, and agricultural processing sectors to do more to bring high commodity prices down — corn and soybean products are important ingredients for those industries. Among other things, these groups have called for the Secretary of Agriculture to announce a penalty-free release of acreage presently under long-term contract in the Conservation Reserve Program (CRP)[28] and for the EPA Administrator to announce a waiver of the Renewable Fuels Standard which mandates an increasing minimum use of biofuels in the national fuel supply.[29]

On April 25, 2008, Texas Governor Rick Perry, in a letter to Stephen Johnson, Administrator of the Environmental Protection Agency (EPA) — the federal agency responsible for administering the RFS — to request that EPA waive 50% of the RFS' ethanol requirements to alleviate their impact on corn prices.[30] However, Governor Perry's request was denied by the EPA.[31]

On May 27, USDA announced that 24 million acres of CRP land could be used in 2008 for a critical feed use (CFU) program of managed haying and grazing following primary bird nesting season.[32] However, a U.S. District Court issued a permanent injunction on July 24 against the CFU except for those who applied before a temporary restraining order issued on July 8.[33]

Flood-related crop production concerns have added to this pressure and have perhaps contributed to the USDA decision on July 7, 2008, to announce that permission is granted in both presidential disaster and contiguous counties to use CRP land for grazing only.[34]

POTENTIAL MARKET IMPLICATIONS DUE TO FLOOD LOSSES

As mentioned earlier, the United States and world markets have experienced tight supplies and record high prices during the first half of 2008.[35] Most long-term forecasts project prices for feed grains and oilseeds — as well as those crops that compete for area with feed grains and oilseeds — to remain at significantly higher levels than experienced during the recent 1998-2006 period.[36] The main factors behind higher long-term prices are projections for a steady rise in global population, accompanied by steady income growth in the world's developing economies, which combine to sustain growth in demand for livestock products and the feedstuffs (e.g., coarse grains and protein meals) needed to produce those products. In addition, the outlook for increased demand for agricultural feedstocks to meet large increases in government biofuel-usage policies, particularly in the United States and the European Union (EU), suggest that demand will increase strongly over the coming decade for corn (the primary feedstock for U.S. ethanol production), and vegetable oils (the primary feedstock for biodiesel production in the United States and the EU).

These long-run forecasts assume normal crop growing conditions and successful harvests. As a result, any deviation from normal growing conditions can be expected to have negative market repercussions and drive prices higher. The potential weather- and flood-related production losses to this year's U.S. corn and soybean crops were unwelcome news to the market and, likely to contribute to higher commodity prices in June. Because the United States plays a dominant role in global corn and soybean markets, U.S. price changes transmit directly to the international marketplace.

In summary, good growing conditions during July and early August of 2008 appear to have moderated initial concerns over potential flood-related crop losses. However, a large portion of the 2008 corn and soybean crops were planted late and, as of early August, remain substantially behind historical development rates. As a result, market analysts suggest that weather problems such as an early freeze could still emerge to lower yield and production prospects. Such concerns are likely contribute to volatile commodity prices, thereby, maintaining pressure on policymakers over concerns about consumer food price inflation, international food aid availability, and the soundness of policy that dedicates commercial agricultural crops to biofuels production, particularly corn used for ethanol.

End Notes

[1] For more information, see CRS Report RL34474, *High Agricultural Commodity Prices: What Are the Issues?*, by Randy Schnepf.
[2] *Crop Production*, National Agricultural Statistics Service (NASS), USDA, August 12, 2008; [http://www.nass.usda.gov/Publications/].
[3] *Crop Progress*, NASS, USDA, August 11, 2008.
[4] *Acreage*, NASS, USDA, June 30, 2008.
[5] "USDA Report Assesses 2008 Corn and Soybean Acreage," USDA News Release, June 30, 2008; at [http://www.nass.usda.gov/Newsroom/2008/06_30_2008.asp].
[6] *Crop Production*, NASS, USDA, August 12, 2008.
[7] Ibid.
[8] *World Agricultural Supply and Demand Estimates (WASDE)*, World Agricultural Outlook Board (WAOB), USDA, July 11, 2008.
[9] USDA *Crop Production* reports are available at [http://www.nass.usda.gov/]; *World Agricultural Supply and Demand Estimates (WASDE)*, at [http://www.usda.gov/oce/commodity/wasde/index.htm].
[10] *World Agricultural Supply and Demand Estimates (WASDE)*, World Agricultural Outlook Board (WAOB), USDA, July 11, 2008.
[11] Note that these calculations by CRS are purely hypothetical. They are available upon request.
[12] Note that these calculations by CRS are purely hypothetical [13] "Memorandum for Reporters and Editors," Iowa Dept. of Agriculture and Land Stewardship, July 1, 2008. Note that Iowa's weather records date back to the early 1870s.
[14] See *Has the best time to plant corn changed?* and *Early planting of soybean is very important*, Integrated Crop Management (ICM), Iowa State University (ISU) Extension, at [http://www.ipm.iastate.edu/ipm/icm/2006/3-13/corntime.html] and [http://www.ipm.iastate.edu/ipm/icm/2007/4-2/earlyplant.html].
[15] *Crop Progress*, NASS, USDA, May 12, 2008.
[16] As cited in "Crop Development Issues, Food Prices and Ethanol Concerns," posted by Keith Good, *FarmPolicy.com*, June 20, 2008.
[17] See *Corn survival in flooded or saturated fields*, and *Planting and replanting scenarios*, ICM, ISU Extension, available at [http://www.ipm.iastate.edu/ipm/icm/2007/4-30/flooded.html] and [http://www.ipm.iastate.edu/ipm/icm/2007/6-4/replant.html].
[18] "Midwest Flooding Affects River, Rail, and Road Traffic," *Grain Transportation Report*, Agricultural Marketing Service, USDA, June 26, 2008. For more information about barge transportation on the Mississippi River, see CRS Report RL32470, *Upper Mississippi - Illinois Waterway Navigation Expansion: An Agricultural Transportation and Environmental Context*, coordinated by Randy Schnepf.
[19] "Grain storage facilities take hit from flooding," by Tim Hoskins, *Minnesota Farm Guide*, July 3, 2008.

[20] Conversation with staff at the Iowa Department of Agriculture and Land Stewardship. Preliminary estimates suggest about 3,500 hogs and no cattle deaths are directly attributable to the June floods.

[21] For more information see CRS Report RL33053, *Federal Stafford Act Disaster Assistance: Presidential Declarations, Eligible Activities, and Funding*, by Keith Bea; CRS Report RL31734, *Federal Disaster Recovery Programs: Brief Summaries*, by Mary Jordan; and CRS Report RL34146, *FEMA's Disaster Declaration Process: A Primer*, by Francis X. McCarthy.

[22] The initial federal disaster declaration was made on May 27, 2008. The final county count for Iowa is available as "Disaster Declaration as of 08/12/2008," FEMA-1763-DR, Iowa, at [http://www.gismaps.fema.gov/2008graphics/dr1763/dec_1763.pdf].

[23] "2008 Federal Disaster Declarations," Federal Emergency Management Agency (FEMA), available at [http://www.fema.gov/news/disasters.fema]. For more information on federal flood response see, "Midwest Flood Response and Recovery," at [http://www.usa.gov/flooding

[24] For more information, see CRS Report RS21212, *Agricultural Disaster Assistance*, by Ralph M. Chite.

[25] For a list of USDA flood-related activities, see "Midwest Flood Response USDA Actions," Release No. 0163.08, updated on July 1, 2008, at [http://www.usda.gov/safety

[26] For more information, see CRS Report RS21212, *Agricultural Disaster Assistance*, by Ralph M. Chite.

[27] For more information, see CRS Report RL34207, *Crop Insurance and Disaster Assistance in the 2008 Farm Bill*, by Ralph M. Chite.

[28] For more information, see CRS Report RS21613, *Conservation Reserve Program: Status and Current Issues*, by Tadlock Cowan.

[29] For more information, see CRS Report RL34265, *Selected Issues Related to an Expansion of the Renewable Fuel Standard (RFS)*, by Brent D. Yacobucci and Tom Capehart.

[30] "Letter to EPA Administrator Stephen Johnson," by Texas Governor Perry, April 25, 2008, at [http://www.governor

[31] "EPA Keeps Biofuels Levels in Place after Considering Texas' Request," EPA News Release, August 7, 2008.

[32] "USDA Announces Crp Permitted Use for Livestock Feed Needs," USDA News Release No. 0137.08, May 27, 2008.

[33] For more information, see CRS Report RS21613, *Conservation Reserve Program: Status and Current Issues*, by Tadlock Cowan.

[34] "USDA Releases CRP Land in Flood Regions for Grazing," Release No. 0179.08, July 7, 2008.

[35] For more information, see CRS Report RL34474, *High Agricultural Commodity Prices: What Are the Issues?*, by Randy Schnepf.

[36] For examples of long-term agricultural forecasts, see *U.S. Baseline Briefing Book*, Food and Agricultural Policy Research Institute, FAPRI-MU Report #03-08, March 2008, at [http://www.fapri.missouri.edu/outreach See also "Agricultural Baseline Projections," Economic Research Service, USDA, at [http://www.ers.usda.gov/Briefing/Baseline/].

INDEX

A

acceptable level of risk, 24, 89, 93
acceptable level of vulnerability, 89
access, 69, 103, 116
accommodations, 30
accountability, 19, 32
accounting, 103, 112, 115
accreditation, 117
acquisitions, 66, 99
ad-hoc appropriations, viii, 97
advisory base flood elevations (ABFE), 116
agencies, 2, 3, 4, 5, 12, 13, 14, 20, 22, 25, 26, 27, 28, 32, 34, 46, 49, 55, 83, 84, 85, 87, 88, 89, 94, 108, 138
aggregation, 108
agricultural market, 124, 137
agriculture, vii, 11, 92, 136
airports, 10
Alaska, 41
annual rate, 113, 115
appropriations, viii, 20, 25, 32, 52, 55, 60, 68, 69, 70, 71, 75, 81, 91, 95, 97, 99, 111, 114, 119
Appropriations Act, 52, 58, 79, 108, 124, 139
assessment, 30, 52, 71, 73, 79, 98, 99, 103, 107, 108, 112, 113, 115
authorities, 65, 72
authority, 15, 18, 42, 49, 52, 72, 86, 88, 110, 139
authorized programs, 138
avoidance, 74
awareness, 4, 28, 31, 53, 90, 99, 115, 116

B

base, 28, 30, 52, 53, 69, 70, 71, 79, 112, 116
base flood elevation (BFE), 112, 116
behaviors, 27
beneficial effect, 59
benefits, vii, 1, 4, 18, 20, 24, 31, 33, 34, 43, 44, 46, 54, 59, 68, 71, 86, 87, 90, 120
biodiesel, 140
biofuel, 140
building code, 30, 55, 75, 102, 113, 119
businesses, 2, 12, 22, 24, 45, 98, 99, 101, 102

C

Cairo, 17
capacity building, 20
capital markets, 114
cash, 115, 125
cash flow, 115
catastrophes, 138
cattle, 142
causation, 118, 122
certification, 5, 14, 109, 117
citizens, 53, 68, 74
City, 8, 10, 30, 47, 48, 109
clarity, 67
classes, 103

144   Index

cleanup, 139
climate, 2, 3, 23, 24, 45, 50, 91, 100, 109
climate change, 23, 24, 50, 100, 109
climate extremes, 23
Clinton Administration, 14, 107
closure, 48
coastal communities, 24, 30
coastal ecosystems, 6
coastal management, 49
combined effect, 114, 128
commerce, 45, 86
commercial, 8, 10, 12, 45, 88, 101, 104, 113, 119, 124, 125, 137, 141
commodity, viii, 123, 124, 137, 139, 140, 141
commodity markets, 124
communication, 13, 72, 109, 119
communities, vii, 2, 11, 12, 14, 19, 21, 24, 27, 30, 32, 45, 51, 53, 54, 55, 57, 58, 59, 61, 63, 68, 69, 70, 72, 74, 76, 84, 85, 89, 90, 92, 94, 100, 101, 102, 104, 113, 116, 119, 120
community, vii, viii, 1, 5, 12, 30, 53, 58, 59, 65, 69, 72, 73, 74, 83, 85, 87, 100, 113, 116, 117, 120, 124, 138
community support, 58
compensation, 102
competitive grant program, 58
competitive process, 52, 57, 60, 69, 70, 71, 73, 79
compliance, 21, 67, 73, 102, 119
conference, 55, 77, 119
configuration, 70
conflict, 32, 118, 119
Congressional Budget Office, 59, 69, 78, 122
congressional hearings, 32, 39
consensus, 63, 94, 113
conservation, 20, 35, 124
Consolidated Appropriations Act, 52, 79
construction, 4, 5, 18, 24, 26, 33, 42, 43, 44, 45, 67, 86, 87, 92, 93, 112, 113, 115, 116, 119
control measures, 90, 110
cooperation, 35, 72, 86
coordination, 4, 13, 14, 15, 19, 49, 67, 117

Corn Belt, viii, 11, 123, 124, 125, 126, 128, 130, 131, 132, 133, 134, 135, 136
cost, 12, 18, 21, 22, 23, 25, 43, 44, 48, 52, 59, 63, 67, 69, 73, 74, 76, 86, 87, 97, 99, 100, 103, 104, 110, 111, 112, 115, 116, 120, 139
cost of living, 23, 116
cost saving, 59
cotton, 132, 133
covering, 22
critical feed use (CFU), 139
critical infrastructure, 7
critical period, 13
criticism, 72
crop, viii, 5, 11, 123, 124, 125, 126, 128, 134, 135, 136, 138, 139, 140
crop insurance, 5, 136, 138
crop production, viii, 123, 124, 125, 126, 128, 140
crops, 124, 125, 126, 132, 133, 135, 136, 140
CRP, 124, 139, 140, 142
CRS report, 4

D

damages, 1, 3, 6, 10, 11, 17, 19, 20, 26, 28, 30, 54, 59, 84, 87, 92, 93, 97, 98, 99, 100, 102, 104, 108, 109, 116, 118
database, 26, 91, 93, 108, 116
deaths, 10, 30, 90, 137, 142
decision makers, 87
Delta, 124, 126, 128
Department of Agriculture, 5, 87, 142
Department of Commerce, 121
Department of Homeland Security, 22, 49, 77, 78, 79, 80, 81, 105, 106, 107, 111, 120, 122, 138
Department of the Interior, 5, 17, 87
destruction, 99
deviation, 140
DHS, 52, 57, 63, 65, 138
digital flood insurance rate maps (DFIRMs), 116
disaster, vii, viii, 1, 2, 3, 5, 8, 10, 12, 20, 21, 22, 24, 25, 30, 32, 48, 50, 52, 53, 54, 55, 57, 58, 59, 65, 68, 69, 71, 72, 74, 75, 76, 78, 81, 83, 85, 87, 88, 94, 95, 97, 99, 100,

Index

101, 102, 103, 104, 108, 109, 110, 112, 116, 119, 120, 124, 136, 138, 139, 140, 뙘142
disaster area, 138
disaster assistance, 2, 3, 5, 12, 21, 22, 30, 85, 88, 95, 109, 110, 116, 120, 138, 139
Disaster Mitigation Act of 2000, viii, 30, 51, 54, 55, 65, 99
disaster relief, viii, 54, 55, 59, 97, 99, 100, 101, 102, 103, 104, 109, 119, 120
Disaster Relief and Emergency Assistance, viii, 51, 55, 99, 138
Disaster Relief Fund, 69, 75, 138
discharges, 7
distribution, 62, 63, 70, 71, 90, 138
drainage, 6, 9, 34, 86
DRF, 69, 75, 76, 81, 138
drinking water, 11
drying, 134, 136

E

earthquakes, 54, 71
economic activity, 50
economic assistance, 98
economic consequences, 11
economic damage, 6, 10, 104
economic damages, 10, 104
economic development, 18, 26, 30, 34, 42, 43, 46
economic growth, 74
economic losses, viii, 10, 83, 104
economics, 101
ecosystem, 17, 43, 44
ecosystem restoration, 43, 44
eligibility criteria, 68
emergency, vii, viii, 1, 2, 3, 5, 12, 13, 22, 29, 47, 52, 54, 62, 67, 68, 69, 72, 73, 83, 85, 86, 87, 88, 90, 95, 97, 99, 116, 119, 124, 138, 139
Emergency Assistance, viii, 51, 55, 99, 138
emergency declarations, 138
emergency disaster loans, 138
emergency management, 54, 67, 69, 70
emergency planning, 22
emergency preparedness, 52

emergency response, vii, viii, 1, 2, 5, 13, 22, 47, 72, 83, 85, 86, 87, 88, 90
encouragement, 6, 45
enforcement, 20, 115, 119
engineering, 15, 28, 47, 71, 117
environment, 26, 34, 102
Environmental Protection Agency, 38, 139
EPA, 41, 139, 142
equipment, 62, 65, 67
erosion, 24, 74, 86, 95, 109, 134, 136
ethanol, 124, 125, 137, 139, 140, 141
EU, 140
European Union, 140
evacuation, 3, 33
Executive Order, 19
exercise, 139
expenditures, 74, 101, 116
expertise, 72
exporter, 124
exposure, vii, 1, 21, 100, 101, 102, 103, 118

F

families, 11, 30, 138
Farm Bill, 142
farm land, 47
farmers, 47, 123, 125, 126, 127, 134, 136, 138, 139
farmland, 10, 12, 20, 139
federal agency, 5, 19, 139
federal aid, 138
federal assistance, 13, 24, 88, 120
Federal Emergency Management Agency, 5, 39, 40, 42, 78, 79, 80, 81, 83, 85, 102, 105, 106, 107, 111, 120, 122, 138, 142
federal funds, 14
federal government, vii, viii, 1, 5, 15, 19, 21, 27, 32, 33, 35, 76, 83, 85, 86, 87, 90, 91, 95, 98, 102, 114, 120
Federal Government, 86
Federal Highway Administration, 38
federal law, vii, 51, 117
federalism, 69
feedstock, 140
feedstuffs, 140

financial, 22, 30, 54, 55, 98, 99, 109, 110, 112, 114, 119
financial condition, 98
financial resources, 110
financial stability, 114
financial support, 55
fire suppression, 62, 68
fiscal deficit, 110
fishing, 30
flexibility, 67, 70, 76, 139
flood hazards, 3, 18, 87, 89, 91, 102, 116
flood insurance study (FIS), 98
flood mitigation, vii, 1, 21, 27, 87, 91, 94
flooding, viii, 1, 2, 3, 5, 6, 7, 8, 10, 11, 16, 20, 21, 22, 23, 24, 25, 34, 45, 47, 53, 61, 75, 83, 84, 85, 87, 88, 89, 90, 91, 94, 97, 98, 99, 100, 101, 104, 106, 109, 113, 115, 116, 117, 118, 123, 126, 128, 134, 136, 137, 138, 141, 142
floodplain management systems, 14, 102
floodwalls, 3, 6, 26, 29, 45, 47, 48, 87, 93
food, 124, 125, 138, 141
force, 14, 53, 104
forecasting, 13, 29
formula, 52, 60, 70, 75, 139
funding, 20, 21, 46, 52, 55, 57, 58, 60, 61, 62, 63, 64, 65, 66, 67, 69, 70, 74, 75, 76, 77, 78, 79, 81, 124, 139
funds, viii, 7, 12, 14, 21, 51, 52, 53, 55, 56, 58, 60, 61, 62, 63, 64, 66, 67, 68, 69, 70, 71, 74, 76, 77, 79, 81, 85, 99, 109, 110, 120, 138

G

GAO, 10, 12, 22, 32, 39, 40, 41, 42, 47, 48, 61, 119, 121, 122
General Accounting Office, 10, 47, 121
global markets, 124
global warming, 100
governments, vii, viii, 1, 2, 3, 5, 51, 53, 61, 65, 67, 69, 74, 75, 77, 83, 85, 86, 87, 89, 90, 108, 116, 120, 138
governor, 142
grant programs, 65, 75

grants, 12, 21, 27, 52, 59, 60, 61, 63, 64, 65, 66, 67, 68, 70, 71, 77, 120
grazing, 124, 139, 140
grids, 91
growth, viii, 23, 51, 74, 135, 136, 140
guidance, 14, 25, 27, 46, 52, 62, 63, 67, 68, 73, 75, 76, 77, 117
guidelines, 18, 43, 44, 93
Gulf Coast, 3, 36, 38, 41
Gulf of Mexico, 30

H

habitat, 12, 44
harvesting, 127
Hazard Mitigation Grant Program (HMGP), 53, 58, 99
hazards, 3, 18, 24, 49, 53, 54, 59, 61, 65, 67, 69, 74, 75, 76, 78, 87, 89, 90, 91, 102, 116, 139
highways, 10, 45, 86, 137
history, 57, 69, 75, 92, 100, 109, 110, 136
homeowners, 21, 98, 103, 109, 117, 118
homes, 2, 3, 8, 10, 12, 22, 30, 45, 53, 101, 108, 109, 119, 120
House, 25, 35, 36, 37, 38, 39, 52, 63, 71, 75, 79, 80, 81, 96, 107, 119, 121
House Committee on Transportation and Infrastructure, 80
housing, 72, 120, 124, 138
Housing and Urban Development, 77, 111, 121
human, viii, 6, 23, 25, 34, 54, 83, 84, 93
Hurricane Andrew, 54
Hurricane Katrina, viii, 2, 3, 4, 6, 22, 24, 25, 26, 31, 37, 38, 40, 41, 50, 59, 78, 80, 83, 84, 88, 90, 91, 92, 94, 95, 106, 109
hurricanes, 3, 23, 24, 54, 98, 100, 101, 103, 106, 107, 108, 110, 116, 118
hybrid, 60, 70

I

ideal, ix, 11, 123, 124, 126
identification, 30, 104, 120
images, 47

Index

improvements, 5, 13, 14, 15, 20, 22, 33, 84, 86, 93
income, 25, 94, 110, 111, 112, 114, 140
individuals, 26, 27, 85, 87, 90, 98, 99, 102, 103, 108, 113, 119, 120, 138
industries, 30, 55, 139
inflation, 124, 125, 141
infrastructure, viii, 2, 3, 7, 10, 11, 15, 18, 19, 22, 24, 30, 31, 44, 45, 83, 84, 85, 86, 87, 89, 91, 93, 94, 109, 137
inspections, 6, 48, 88
institutions, 20
integrity, 7, 15, 110
investment, 2, 3, 43, 84, 86, 87, 93, 94
investments, vii, viii, 1, 2, 3, 5, 6, 7, 14, 20, 22, 27, 46, 76, 84, 89, 90, 91, 92, 97, 102
Iowa, 7, 8, 10, 11, 17, 46, 47, 48, 99, 104, 105, 125, 127, 128, 129, 130, 132, 133, 134, 136, 137, 138, 141, 142
islands, 24
isolation, 6
issues, viii, 3, 13, 19, 32, 39, 42, 51, 54, 62, 65, 84, 85, 91, 93, 96, 98, 99, 100, 102, 107, 117

J

Jordan, 142
jurisdiction, 65, 96, 120
justification, 60, 61

K

kill, 136

L

labeling, 107
laws, 24, 35, 74
lead, 13, 102, 116
leadership, 4, 46, 74
legislation, 3, 4, 16, 21, 25, 32, 49, 52, 55, 57, 62, 67, 68, 69, 71, 74, 77, 110, 118, 119
legislative proposals, 98, 99
levees, vii, 1, 3, 4, 5, 6, 7, 8, 10, 12, 13, 14, 15, 17, 20, 22, 26, 27, 28, 43, 45, 46, 47, 48, 53, 67, 68, 85, 87, 88, 89, 90, 91, 92, 93, 98, 99, 100, 101, 102, 103, 108, 109, 110, 113, 115, 117, 119, 136
light, 69, 124
livestock, 125, 127, 137, 139, 140
loans, 19, 113, 115, 120, 138
local authorities, 72
local community, 120
local government, viii, 2, 3, 4, 5, 51, 53, 57, 61, 65, 67, 69, 72, 73, 74, 75, 77, 83, 85, 86, 87, 89, 90, 92, 108, 117, 120, 138

M

machinery, 58
magnitude, 2, 6, 9, 10, 46, 89, 100, 104
majority, 2, 17, 46, 52, 59, 66, 75, 77, 104
man, 12
management, vii, 2, 3, 4, 6, 14, 15, 16, 17, 18, 19, 20, 21, 25, 27, 44, 49, 53, 54, 64, 66, 67, 69, 70, 74, 84, 85, 89, 90, 91, 95, 99, 101, 102, 103, 104, 107, 108, 109, 112, 113, 120
mapping, 21, 22, 67, 85, 89, 103, 104, 107, 112, 113, 115, 120
market penetration, 103, 119
marketing, 122, 137
measurements, 13
membership, 17, 103
methodology, 46, 71, 114, 134
Mexico, 30
Midwest flooding, 2, 116
Midwest Interagency Levee Task Force, 14
mission, 17, 32
missions, 32
Mississippi River, 2, 4, 7, 8, 9, 10, 11, 12, 15, 16, 17, 18, 42, 44, 45, 46, 47, 48, 49, 50, 83, 86, 87, 95, 97, 99, 100, 101, 121, 136, 137, 141
Missouri, 8, 9, 10, 11, 16, 17, 18, 39, 44, 46, 99, 105, 125, 127, 128, 129, 131, 132, 133, 134, 136
modernization, 14, 117
modifications, 92
moral hazard, 103
mortality, 138

Index

N

NAP, 138
National Environmental Policy Act (NEPA), 79
National Flood Insurance Program (NFIP), viii, 5, 74, 76, 88, 97, 98, 99
National Oceanic and Atmospheric Administration (NOAA), 100
National Research Council, 22, 91, 95
National Response Framework, 88
national strategy, 27
natural disaster, viii, 53, 65, 88, 97, 101, 102, 119, 138, 139
natural disasters, 53, 88, 101, 139
natural hazards, 54, 74
nonprofit organizations, 138
North America, 121

O

Office of Management and Budget, 14, 115
officials, 30, 45, 53, 55, 69, 72, 73, 74, 117, 128
oil, 30
operations, 3, 88, 91, 95
opportunities, 13, 16, 20, 43, 44
outreach, 142
oversight, 18, 20, 32, 46, 69, 110, 118, 119, 120
ownership, 23, 116

P

Pacific, 122
participants, 13, 65, 69
peer group, 70
peer review, 63, 67
penalties, 119
plants, 10, 45, 126, 136, 137
policy, iv, vii, 2, 3, 4, 12, 14, 15, 16, 18, 19, 20, 21, 22, 25, 26, 27, 33, 34, 35, 54, 75, 76, 77, 85, 86, 91, 92, 93, 94, 96, 98, 99, 100, 101, 110, 112, 119, 122, 124, 125, 141
policy choice, 27
policy issues, 98, 99, 100
policymakers, 100, 102, 112, 124, 139, 141
pools, 118
population, 2, 3, 6, 12, 17, 19, 20, 22, 23, 45, 54, 70, 91, 100, 140
power plants, 45
precedent, 87
precipitation, 3, 7, 12, 23, 45, 47, 100
pre-disaster conditions, 48, 120
Pre-Disaster Mitigation (PDM), vii, 51, 52, 56, 57
preparedness, 5, 29, 52, 53, 62, 65, 67, 68, 88, 94, 95
preservation, 67
President, 48, 49, 57, 70, 84, 93, 108, 110, 119, 121, 138
President Clinton, 57
prevention, 33, 119
price changes, 140
probability, 3, 6, 24, 87, 88, 89, 93
producers, 125, 127, 138
program administration, 61, 64
program staff, 67, 76
project, 5, 18, 19, 25, 33, 35, 44, 46, 48, 53, 61, 64, 66, 67, 68, 69, 72, 73, 74, 86, 87, 92, 140
Project Impact, vii, viii, 51, 52, 53, 56, 57, 58, 69
project sponsors, 5, 86
property rights, 2, 25
protected areas, 10
protection, vii, 1, 6, 7, 10, 12, 14, 15, 18, 20, 24, 25, 26, 43, 44, 45, 48, 85, 87, 88, 89, 90, 91, 92, 94, 95, 107, 108, 109, 117, 119
provisional accredited levee (PAL), 117
public awareness, 99
public concern, 24, 93
public concerns, 93
public health, 93
public safety, 21, 30, 110

Q

quality standards, 23

Index

R

rainfall, 9, 13, 47, 97, 99, 108, 109, 134, 135
recognition, 60
recommendations, iv, 4, 16, 17, 18, 19, 20, 21, 22, 25, 26, 27, 43, 84, 93, 95, 101
reconstruction, 14, 18, 44, 108
recovery, vii, viii, 1, 3, 5, 11, 14, 22, 29, 30, 32, 57, 69, 72, 81, 119, 123
recovery plan, 30
recreation, 45
recurrence, 47, 100
redevelopment, 12, 29
reform, 112, 119
Reform, 21, 25, 36, 38, 103, 118, 119, 122
regulations, 12, 29, 30, 73, 74, 112, 116
rehabilitation, 17, 24, 27, 33, 43, 54, 87, 88, 139
Rehabilitation and Inspection Program, 5, 13, 14, 88
rehabilitation program, 17
reinsurance, 114, 115
reliability, 7, 13, 26, 85, 88, 90, 91, 93, 94, 115
relief, viii, 54, 55, 59, 97, 99, 100, 101, 102, 103, 104, 109, 119, 120
Renewable Fuel Standard, 142
Renewable Fuels Association, 137
repair, 5, 13, 14, 48, 57, 87, 88, 99
requirements, 13, 14, 22, 27, 48, 89, 93, 98, 102, 103, 112, 113, 115, 139
researchers, 55
resolution, 32
resource management, 15, 49
resources, viii, 15, 16, 17, 19, 21, 24, 25, 26, 29, 30, 34, 35, 51, 55, 58, 62, 63, 65, 72, 84, 86, 87, 90, 92, 93, 110, 124, 138
response, vii, viii, 1, 2, 3, 5, 6, 12, 13, 22, 47, 53, 55, 57, 65, 69, 72, 81, 83, 84, 85, 86, 87, 88, 90, 94, 95, 97, 100, 102, 119, 126, 138, 139, 142
restoration, 14, 33, 43, 44, 48, 139
revenue, 120, 139
RFS, 139, 142
risk assessment, 71, 103, 107, 112, 113

risk management, vii, 6, 25, 27, 49, 85, 90, 99, 109
risks, 2, 3, 14, 21, 23, 24, 25, 30, 31, 50, 54, 84, 87, 88, 91, 98, 100, 101, 109, 110, 114
river basins, 100
rowing, ix, 11, 123, 124, 126, 140

S

SACE, 101
safe room, 53, 65
safety, 4, 21, 25, 26, 27, 28, 30, 34, 84, 88, 93, 95, 110, 142
scientific knowledge, 91
scope, viii, 17, 19, 45, 51, 55
sea level, 23
sea-level, 91
sea-level rise, 91
Secretary of Agriculture, 139
security, 33, 86, 90
seed, 136
Senate, 25, 35, 36, 37, 38, 39, 48, 49, 52, 60, 78, 95, 96, 119, 121
shape, vii, viii, 1, 5, 25, 83
shoreline, 95
shortfall, 87, 114
showing, 76
simulations, 47
social security, 33, 86
South Dakota, 8, 105, 125, 128, 129, 131, 132, 133
soybeans, viii, 11, 123, 124, 134, 135, 136, 137
spending, 24, 55, 59, 60, 62, 65, 66, 70, 74
Spring, 134
stability, 114
Stafford Act, 52, 56, 58, 60, 65, 69, 73, 78, 95, 138, 142
stakeholders, 18, 46, 48, 90
state, viii, 2, 4, 5, 15, 16, 18, 19, 21, 22, 24, 27, 30, 31, 32, 33, 35, 45, 48, 49, 51, 52, 53, 54, 57, 58, 60, 61, 63, 65, 66, 67, 69, 70, 71, 72, 73, 75, 77, 83, 85, 86, 87, 89, 92, 94, 99, 108, 117, 118, 120, 125, 128, 135, 136, 138

Index

states, 1, 2, 3, 9, 12, 15, 18, 19, 21, 24, 27, 35, 42, 46, 48, 50, 52, 58, 60, 61, 63, 64, 65, 69, 71, 73, 74, 75, 76, 90, 93, 94, 97, 99, 100, 104, 105, 118, 119, 125, 127, 128, 134
statutes, 52, 55
statutory authority, 49
steel, 47
storage, 6, 45, 137, 141
storm shutters, 68
storms, 1, 2, 3, 23, 24, 47, 84, 88, 102, 134, 138
stress, 135
structure, 2, 26, 89, 102, 112
suppression, 62, 68
survival, 141
synchronize, 9

T

T&I, 35, 36, 38, 39
target, 114
taxpayers, 98, 102
technical assistance, 64, 66
technical support, 46
techniques, 114
technology, 71, 91
temporary housing, 72, 120
Tennessee Valley Authority, 5, 50, 87
tensions, 32
territory, 63
terrorism, 65
threats, 69, 91
time constraints, 17
tracks, 91
trade, 124
training, 28, 118
transmission, 116
transportation, 10, 11, 18, 22, 44, 137, 141
transportation infrastructure, 18, 44, 137
Treasury, 104, 110, 111, 113, 115, 120
treatment, 10, 20, 29, 45
triggers, 89

U

U.S. Army Corps of Engineers, 5, 28, 41, 42, 47, 53, 83, 85, 101, 121
U.S. Geological Survey, 47, 48, 120
U.S. policy, 34
U.S. Treasury, 110, 111, 115
underwriting, 102, 109, 116, 122
uniform, 108
uninsured, 97, 99, 103, 109, 110, 116, 120
United, iv, vii, viii, 1, 5, 16, 23, 26, 33, 47, 49, 83, 85, 86, 89, 90, 98, 100, 104, 120, 121, 124, 130, 131, 132, 133, 140
United States, iv, vii, viii, 1, 5, 16, 23, 26, 33, 47, 49, 83, 85, 86, 89, 90, 98, 100, 104, 120, 121, 124, 130, 131, 132, 133, 140
updating, 23, 103, 115, 116
Upper Mississippi River Comprehensive Plan (UMRCP), 4, 16, 42
urban, 3, 6, 7, 15, 24, 43, 45, 47, 48, 89, 90, 91, 92, 94
urban areas, 7, 43, 45, 47, 48
urban population, 3, 24, 92, 94
urbanization, 100
USDA, 40, 123, 124, 126, 127, 128, 130, 131, 132, 133, 134, 135, 137, 138, 139, 140, 141, 142

V

valuation, 18, 42
variations, 89, 113
vegetable oil, 140
vegetation, 50, 53
velocity, 6, 15
victims, 99, 102, 109, 119
vulnerability, vii, 2, 3, 17, 19, 24, 51, 84, 88, 89, 90, 93, 94, 102, 108, 110, 116, 136

W

waiver, 139
War on Terror, 108
warning systems, 65, 67, 68
Washington, 49, 73, 77, 78, 81, 95, 120, 121, 122

Index

wastewater, 45
water, 5, 7, 10, 11, 15, 16, 17, 19, 21, 23, 25, 26, 32, 34, 35, 45, 47, 48, 84, 87, 93, 120, 122, 134, 136
water resources, 15, 16, 17, 19, 21, 25, 26, 34, 35, 84, 87, 93
Water Resources Development Act (WRDA), 4, 84
watershed, 6, 17, 34, 44, 136
waterways, 86

wealth, 23, 94
welfare, 20, 86, 93
Wisconsin, 7, 8, 11, 99, 105, 125, 127, 129, 131, 132, 133, 134, 135, 136, 138

Y

yield, viii, 123, 124, 126, 127, 128, 134, 135, 136, 140